MIND STATES
An Introduction to Light & Sound Technology

Michael Landgraf

 Published by Little Minnie's House. Printed in the United States. All inquiries pertaining to this book should contact: miccar@ix.netcom.com

INTRODUCTION

I purchased my first light and sound unit in 1986. Five years later, light and sound became my profession. It has been an incredible journey. You are about to learn what I have seen.

Having graduated from college as a cultural anthropologist, my main interest has always been one of participant-observation. Getting involved, then stepping back and watching what happens. This is how I approached light and sound, why I have volunteered hundreds of hours and donated dozens of instruments. I have sponsored programs that have benefitted teachers and students, doctors and patients, businesses and athletes. Everyday people.

This is not a textbook and certainly not my thesis. It is a little book that will show you how to use a light and sound instrument. Very simple and very basic. It will tell you all you need to get started and where to go to learn more. It is a book that should have been written years ago.

I am a light and sound dealer who works with professionals that did not use light and sound instrumentation until we met. It has been very enjoyable introducing professionals to a new technology. Watching skepticism turn into excitement. Seeing new applications net positive results. What we have observed has sometimes been completely different to what many of the leaders in this field have led us to believe will occur. It is not my intention to upset anyone, certainly not the leaders behind light and sound technology, but facts are facts. You, the user of light and sound equipment, deserve to know these facts.

Why these differences have occurred is very simple. No two people think exactly alike. The mental, emotional and physical sensations that one individual experiences from a light and sound session will be similar, but not the same as the next individual. After all, how can we expect the response to light and sound from an individual who has AIDS to be the same as, for example, those experienced by an emotionally upset police officer?

In the first section you will learn what a light and sound (and sound only) instrument is. You will also learn the general characteristics of each brainwave state and how, by utilizing light and sound frequency stimulation, these different brainwave states can be accessed. Also, you will gain an understanding of how frequencies and brainwaves are associated and whether or not this technology is safe for you to use.

I began sponsoring learning/relaxation programs in 1991. Working with professionals in the medical, educational, business and law enforcement communities, we utilized light and sound frequencies in an effort to determine their impact upon targeted brainwave states. How did we use light and sound? Were the sessions we used effective? Did these sessions elicit positive results? In the second section, I will share the answers with you. How professionals use light and sound stimulation to access beta, alpha, theta and delta brainwave states can be your guide as you begin your journey towards consciousness exploration and personal relaxation.

How do you choose a light and sound (or sound only) instrument? What type of system would best suit your needs? Section three will help you in answering these questions. Is the system easy to use? Durable? What about eye frames and headphones? What types of light and sound generation are these instruments capable of producing? And how does tactile (physical vibrations) stimulation effect a light and sound experience? These are very important questions you need to answer when you purchase a light and sound system. This section will help you make the right decision.

The next section will show you how simple it is to use light and sound (and sound only) instruments and why you have no reason to be overwhelmed by this technology. Quick starting your unit, along with helpful hints, will enable you to begin your light and sound experience immediately. You will see what adding music or spoken word can do. You will learn what built-in and downloadable sessions are and how, if these sessions do not meet a specific need, you can design your own light and sound session to meet that need.

Section five will lay the foundation for your journey into light and sound technology. There is a 30 day program that will start you in the right direction. It is a flexible guide that can easily fit into anyone's busy schedule. And there is a daily journal for you to write down your session experiences and sensations. This personal daily journal will allow you in one month to look back and compare the mental, emotional and physical sensations that you will be experiencing with how you are feeling today. Use it! You will be glad you did.

This book ends with some thoughts I wish to share with you and my clients. Light and sound technology has grown immensely since I began my journey, but roadblocks to general acceptance still remain. I will leave you with some questions and illustrate their importance. It is my hope that you share your light and sound experience with others, so these roadblocks will gradually come down. To be replaced with an understanding that this technology is indeed effective, and that it is extremely beneficial.

A suggested reading list is included for assisting you in the areas that you would like to learn more about and also a glossary for all those new words you are about to come across.

My goal for writing this book is to provide you with a basic understanding of how to use your light and sound (and sound only) instrument safely and effectively, what you should expect from light and sound frequency stimulation and how you can maximize your session experiences.

Electrical Brainwave Frequency Patterns
(Illustrating the four general states of mind)

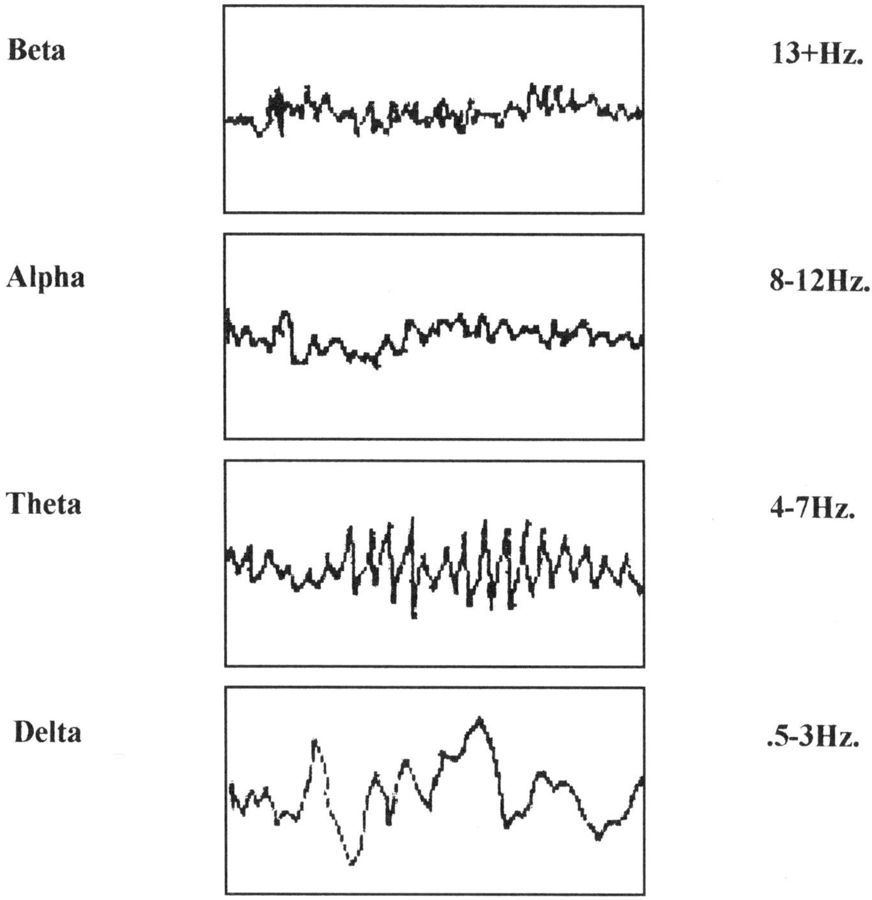

I. WELCOME TO LIGHT & SOUND TECHNOLOGY!

A. What Is A Light & Sound Instrument?

A light and sound instrument is a mind entrainment tool. The method by which this entrainment occurs is known as the Frequency Following Response (FFR). Through the use of audio (headphones) and visual (eye frames with LEDs) stimulation, listeners are gently guided into specific states of mind. Each audio beat and light pulse is a specific frequency. Our minds "think" in terms of frequency. Brainwaves change frequencies based on neural activity within the brain, be it by hearing, touch, smell, vision and/or taste. These senses respond to activity from the environment and transmits that information to the brain via electrical signals. Hearing and vision are considered favorable senses for affecting brainwaves safely. By presenting these beats and pulses to the brain, within a few minutes, the brain begins to mimic or follow the same frequencies as the stimuli (the beats and pulses). This process is referred to as entrainment. In essence, these instruments speak to the mind in it's own language- the language of frequency.

The Frequency Following Response that the mind experiences through light and sound stimulation provides a focus for the brain. This focus safely quells any internal dialogue or "brain chatter", and enables the listener to reach more quickly and effectively the optimum state of mind targeted for the task at hand (i.e. energizing, learning, relaxing, etc.).

So a light and sound instrument can entrain your mind to experience different levels of your consciousness. As such it is a tool, but it is a very enjoyable tool. For a whole new musical experience combine your favorite cassette or CD with a light and sound session. You can actually see, as well as hear, the music as it travels along your brainwaves. It is also for exercising your mind. After all, your brain is like a large muscle. Use light and sound to train your brain to run faster and more efficiently; to show you how to find and access your "zone" and to help clean up that mental fuzz that occasionally interrupts our mental clarity.

Humanity has been aware of how outside stimulation can affect the mind long before recorded history. We can share that same awareness by sitting around a campfire. Watching the flickering flames creates a soothing, calming sensation within the mind. This same affect is also experienced with rhythmic drumbeats. The continuous, repetitive beats tend to have a calming affect, while faster repetitions create a more energetic sensation. Within the twentieth century, researchers have developed this awareness into highly beneficial instrumentation.

Although scientists knew flickering light and audio beats affected our frame of mind, they were unable to prove that an interaction existed. Then in 1924, Hans Burger, a German psychiatrist, published pictures depicting electrical activity of the human brain. From the discovery of these "wavy" lines emerged a new scientific field called electroencephalography.

These wavy lines are simple to visualize. Just picture a series of peaks and valleys, much like the skyline of a mountain range. When you are asleep, those peaks and valleys are at both their highest and lowest points and appear to be spread out. As you gradually awaken and become more alert, the peaks and valleys begin to shrink in size and are spaced closer together. Then when you are in your normal waking state, this "mountain range" is at it's smallest level, with very little space separating the peaks from the valleys.

During the 1940's several researchers, including W. Gray Walter, utilized powerful electronic strobes with new versions of EEG instrumentation to alter brainwave activity, producing states of profound relaxation and imagery. In 1949, brainwave signals were brought to the screen with the invention of the Tuposcope. This breakthrough allowed the tracking of brainwave patterns (Beta, Alpha, Theta and Delta). Within five years, hundreds of hospitals were compiling EEG's.

In the 1950's and 1960's, research on Zen and Yoga meditators showed a predominance of alpha and theta waves during meditation. These findings were made through the data collected by researchers including M.A. Wanger of the University of California at Los Angeles; B.K. Bagchi of the University of Michigan School of Medicine; and B.K. Anand of the All India Institute of Medical Sciences in New Delhi (Ditson-Sommer

1995).

During the late 1960's and into the 1970's interest in producing altered states without chemicals increased. Alpha EEG feedback by Dr. Joe Kamiya at Langley-Porter Neuropsychiatric Institute in San Francisco aided in initiating the age of biofeedback. Others, including Jack Schwarz and Richard Townsend, continued their investigations of rhythmic light and sound combinations, with emphasis on hemispheric brain synchronization and EEG entrainment (Budzynski 1991). It was during the decade of the 1970's that the first patent was obtained for a light and sound instrument. This occurred in 1974, with the patent obtained by Seymour Charas. He was a scientist at New York City College and for reasons unknown, never put his device into production (Hutchison 1990).

In the 1980's there were major breakthroughs in microelectronics. From this came the development of instruments that contained various programs of sound frequency changes and light intensity modulation. In a study by Marchal Gilula, M.D., of Life Energies Research Institute of Coconut Grove, Florida, clinical research on Multiple Afferent Sensory Stimulation (MASS) was conducted. His research found that light and sound (MASS) equipment elicited an 80% subject response with deep states of mind and body relaxation (Ditson-Sommer 1995).

For the past several years, clinical devices utilizing light and sound stimulation have been successfully used in relaxation and learning enhancement. For the general public, small portable units became available. Still, this technology is so new that the majority of us do not know how to use it. Those that do know are using it for commercial purposes, ranging from audio/video cassettes and CD's to seminars.

B. How Are Frequencies Associated With Brainwaves?

Brainwaves are the electrical wave patterns generated in every person's brain. These waves vary according to level of consciousness, subconsciousness and unconsciousness and are characterized by four distinct types of brainwaves. Each of these electrical wave patterns have distinctly different ways of perceiving, processing, learning and knowing information (Hoiberg 1989). All of these brain waves are produced at all times. However, a predominance of a specific desired brainwave state can be created at will, which allows a person to potentialize his or her capabilities towards achieving human excellence.

Brainwave frequencies are described in terms of hertz (Hz), or cycles per second, which are measured by an electroencephalogram (EEG). The four general categories of brainwave frequencies and their main characteristics are:

Gamma brainwave frequencies are a high frequency pattern beginning at thirty-five hertz. There has been some evidence that gamma frequencies are associated with states of peak performance. Most light and sound instruments have the capability to generate gamma frequencies. While in this brainwave state, sensations are centered on being mentally, emotionally and physically "charged" or extremely energized. These frequencies are the highest known brainwave patterns, but are considered to be part of the beta frequency category.

Beta brainwaves are the next highest frequency pattern, ranging from thirteen to thirty-four hertz, and can be separated into three sub-categories. The first is high beta, ranging from nineteen to thirty-four hertz. When high beta is the dominant brainwave state, anxiety and stress are most likely to occur. The second sub-category is mid beta, with frequencies ranging from fifteen to eighteen hertz. Mid beta is characterized by action, with focus on external surroundings. The third is SMR beta (Sensorimotor Rhythm), ranging from thirteen to fifteen hertz. While in this state, focus is also on external surroundings, but the individual is more relaxed than in mid beta. Within SMR is fourteen hertz, which is gaining attention from researchers and clinicians utilizing EEG/neurofeedback instrumentation. The true benefit of creating fourteen hertz in humans has yet to be determined, but a recent study at the University of California at Los Angeles found that felines can produce fourteen hertz at will, while humans cannot.

Generally speaking, beta brainwaves are characterized by logical, analytical and intellectual thinking, verbal communication and awareness of surroundings.

Alpha brainwaves are a frequency pattern ranging from eight to twelve hertz. They most commonly occur when we are calm and relaxed, yet mentally alert. These brainwaves are also present during daydreaming.

Theta brainwaves range from four to seven hertz, characterized by being deeply relaxed and inwardly focused. This brainwave state is also associated with rapid learning and the assimilation of new information with high retention, heightened motivation to activate goals, bursts of creativity, insight and new behavior patterns.

Delta brainwaves range from five-tenths to three hertz and are associated with being extremely relaxed, characterized by sleep.

Researchers have proven that brainwave frequencies determine what brainwave state is being experienced at any given time (Budzynski 1991). By measuring the production of brainwaves through the utilization of EEG (electroencephalogram) equipment, the frequencies that produce these various brain states can be tracked on a computer monitor. As mentioned earlier, these frequencies are generated in every person's brain, and are the result of outside stimulation that has been passed to the brain via electrical signals from our different senses. Light and sound instruments stimulate the neural activity of our vision and hearing, and can safely guide our minds to produce specific frequencies that are associated with the four general brainwave categories (Hutchison 1992).

C. Are Light & Sound Instruments For Everyone?

You must need to know if light and sound is safe for you to use. Generally speaking, the answer is yes, when used properly and with understanding. For example, do not listen to a session half way through then go out and drive a car. Allow the session to end, bringing your awareness back to your normal low beta state, before performing tasks involving heavy machinery. Do not use light and sound without professional medical supervision if you have epilepsy, heart trouble, latent seizure tendencies or are photosensitive. Some sessions promote the emergence of suppressed events and issues. If you have never had therapy or have trouble handling these issues, be very careful in choosing sessions.

Before you purchase a light and sound instrument, ask to see the owners manual. Review it. Give the section that contains the at-risk warning special attention. Be sure to read the disclaimer. Just like every other professional, manufacturers must take precautions to protect their right to produce these instruments. But you have the right to protect yourself. Read what they say regarding those who could be at risk of possible adverse effects from pulsed light stimulation. If you find yourself as an at risk individual, seek advice from your physician before using a pulsed light device.

Remember: everyone is different. No two people think exactly alike. The effects from using light and sound equipment will vary from individual to individual. So play it safe. If you have questions, get the answers before experiencing light and sound sessions.

Disclaimer: Please read carefully!

The information contained in this book is intended for individuals who will be using light and sound and/or sound only instrumentation. Both the author and publisher accept no responsibility or liability for inaccurate, incomplete or misleading information that may be contained in this book.

Light and sound instrumentation is not intended as a treatment for any mental, emotional or physical ailment or condition. Persons with a history of epilepsy or any form of brain seizure should not use a pulsed light device without professional medical supervision. Latent seizure tendencies, brain injuries, toxemia, drug abuse and unspecified other conditions, known or unknown to the user, are conditions which may produce brain seizures. The author and publisher accepts no responsibility or liability for seizures or other negative effects that may occur to individuals utilizing light and sound and/or sound only instruments.

The information contained in this book is not intended to replace any information or advice given by professionals in the medical community.

The author and publisher of this book do not claim that the information contained on these pages are sufficient to provide absolute protection against the risks of exposing individuals to light and sound and/or sound only technology.

You, the reader of this book, understand that this technology produces visual and auditory stimuli to naturally affect brainwave activity. You also understand that both the author and publisher do not make any claims as to this technology's suitability for all applications or users, because the medical or mental state of all users cannot be preassessed. Each user may be affected differently when utilizing light and sound and/or sound only, due to individual emotional, mental and physical characteristics. You, the reader, assume all risks in connection with using any type of light and sound and/or sound only product. You are advised by the author and publisher to seek guidance from qualified medical professionals regarding any and all risks from using light and sound and/or sound only instruments.

Are you an individual who would be at risk by using light and sound technology? Less than 2% of the U.S. population are likely to ever experience a photoconvulsive seizure from pulsed light stimulation (Newmark & Penry in Niedermeyer 1982).

As of July, 1997, the only true method of determining whether or not you are prone to a seizure of this type is to be screened by a qualified technician using EEG analysis prior to using light and sound equipment. However, many epileptics may not experience any adverse effects from light and sound because they are not photoconvulsive (Stevens in Niedermeyer 1982). Still, it is wise to seek professional medical advice prior to using light and sound equipment if you have a history of epileptic or photoepileptic tendencies. Even though medical professionals can identify most of us who have these tendencies by reviewing our individual family histories, some people slip through the screening process undetected who are actually at risk. Why? Because some may be totally asymptomatic or have forgotten that when they were young (commonly between the ages of six and fifteen) they did have a seizure.

The following list of questions are very important for you to ask yourself prior to using light and sound, but collectively should not be construed as a reliable screening procedure. Have you ever had periods of blanking-out or have suffered from any form of convulsive seizure? Have flashing lights ever made you nauseous or uncomfortable? Has anyone in your family ever been prone to seizures? Are you an individual in the beginning stages of drug or alcohol withdrawal? Are you suffering from severe medical disorders? Do you suffer from any type of mental illness? Do you have a pacemaker or are you pregnant? If you answered yes to any of these questions you should consider yourself at risk for possible adverse effects from utilizing pulsed light stimulation.

If you find you are unable to use light and sound because of possible adverse effects, there is an alternative. Sound only. This utilization of audio frequency sessions has proven quite successful in each one of my learning and relaxation programs.

Light and sound instruments give you the ability to control your mental states. Because each person has their own psychological makeup, no one can guarantee you will experience positive feelings after every session. Using light and sound is a personal choice, and the better informed you are, the better experience you will have.

Before ending this discussion concerning the possibility of adverse effects, I feel it is important to address one issue that a majority of my clients and associates have, at one time or another, asked about. If you consume alcohol or drugs (legal or otherwise) before experiencing a light and sound session, many researchers contend that you will automatically have a brain seizure. This is not true. Chances are you will, but it is not automatic. After seven years of sponsoring light and sound programs, not once has there been an adverse reaction to light and sound from any of the participants of these programs. Many were on prescribed medications, while others were confronting their drug or alcohol dependency. Others combined alcohol with light and sound during social gatherings and I repeat, not once did anyone experience any adverse effects.

Here is one example, in a social setting, where individuals consumed alcohol prior to experiencing light and sound. A few years back, representatives of a major international music company asked for my assistance at a party they were having to publicize a new rock group. It was held at a very respectable nightclub on Santa Monica Boulevard. The news media was well represented at this function.

During the course of the festivities, I performed over seventy light and sound sessions on curious individuals who wanted to experience what light and sound was all about. I can not and will not say drugs of any kind were used by any of the attendees. I can say alcohol was consumed. The majority of those experiencing light and sound sessions did indulge. In a way it was almost comical. I asked each person before setting them up on light and sound if they were under the influence of either alcohol or drugs. Of course they all answered no, even with red eyes and silly grins that told me otherwise. Now if I believed everything that has been written about this combination, I never would have agreed to hook anyone up. But no one suffered any adverse effects unless you term having a great time adverse.

Here is an example where prescribed medications could possibly lead to adverse effects when combined with light and sound, yet none were experienced. A few years back, a client started EEG/neurofeedback training after being diagnosed with Chronic Fatigue Syndrome. Over the course of the previous ten months, she had been prescribed over sixty different medications to combat what at that time was an undiagnosed illness. Yes, you read that correctly. Over sixty. Even in this day and age, I guess overkill is still applicable. She was gradually weaned from the majority of the medications and began utilizing light and sound to help her cope with daily stressors while she did her neurofeedback training. Although she was still taking some of her prescribed medications while experiencing light and sound, not once did she suffer any adverse effects from the combination.

I am certainly not advocating the use of alcohol or drugs. And I do not advocate their use with light and sound equipment. What I am saying is people have been known to use both simultaneously and not experience any adverse effects. That is a fact.

II. ACCESSING BRAIN STATES WITH LIGHT & SOUND AND SOUND ONLY INSTRUMENTS

A. Benefitting From Light & Sound Technology.

I began sponsoring learning/relaxation programs in 1991. Participants throughout the greater Los Angeles area have utilized this technology in a variety of applications. Working with professionals in the medical, educational, business and law enforcement communities, we have been able to observe how specific light and sound/sound only sessions directly influence people. The participants came from a variety of backgrounds, their ages ranged from pre-school to the elderly and their health varied from physically fit to the terminally ill.

I have witnessed some fascinating things when people have used light and sound. Before venturing into how others use their instruments on a general level, I would like to share with you two experiences that have made a big impression on me and hopefully will be "food for thought" for you, in terms of the potential and possibilities of light and sound technology.

The experience with Darby Hewitt happened in the spring and summer of 1992. Darby was a head nurse at a large Hollywood, California hospital until she started suffering from the spastic type of Muscular Sclerosis. In a short period of time, she became confined to a hospital bed and wheelchair. Her left arm and hand, along with both legs, were paralyzed.

Because of the M.S. and the medications she took, there was a strong possibility of seizures occurring should she utilize light and sound stimulation. So she sought and received medical clearance from her physician. She then started utilizing an I.Q. Tutor by InnerQuest. She used her unit frequently, often doing multiple sessions daily. Darby never did experience any seizure activity. Within one month after beginning her sessions, Darby experienced less muscle cramping, a decrease in muscle spasms, the fingers of her left hand began to move on command (after more than two years of paralysis) and she gained an overall sense of well-being. Like her doctor, her nurse and her physical therapist, both Darby and I were totally amazed. Incidentally, Darby is another good example of combining strong medications with light and sound and not experiencing any adverse effects.

The other experience happened in the winter of 1997. A friend of mine, Chris Nolind, had lost the majority of feeling in both legs. Chris called and asked if I would help him deal with the stress this was causing him. Although common sense told me otherwise, I agreed because he is my friend.

Under no circumstances did we approach this in a "professional medical" manner. Instead, all we wanted was to help him relax. Now Chris is a prime example of how light and sound can effect everyone differently. Normally alpha frequencies will create a feeling of calmness, but it is different for Chris. Having done sessions in the past where his base rates (heart rate, left and right brain states, left and right GSR and skin temperature) were taken before and after each session, we agreed to utilize higher frequencies. Gamma to be exact.

When Chris arrived for his first session, he was walking with the aid of a cane. That first session, as with the five that followed, consisted of gamma light, sound and tactile stimulation combined with motion (tactile stimulation and motion are part of Integrative Motion System's bed that I've been utilizing since August of 1995).

The gamma session Chris experienced was custom designed specifically for him. It consisted of gentle ramping between 37Hz and 42Hz with a time length of 30 minutes. The instrument we used was a Photosonix Nova Pro. Possessing two outlets for both light and sound, we were able to connect the eye frames, headphones and transducers (bass speakers) of the IMS bed directly into this one unit, allowing total control over the light, sound and tactile stimulation.

This high frequency, multiple stimulation method was based on a little known technique created by David McLaughlin and referred to as the "Principle of Thwack". And that is exactly what it does. After the first session ended Chris stood without the aid of his cane. In fact, he stopped using it entirely. After the third session, Chris went jogging. After the fourth, he took his girlfriend dancing. And after completing all six sessions, Chris was back to his old self again.

It is my opinion that both Darby and Chris succeeded because they were able to access the areas of their minds that gave them the power to do so. The light and sound units were only vehicles used to get to that "place". Now if they can overcome the obstacles they faced by using light and sound stimulation, imagine what your instrument can do for you. The possibilities are truly endless.

While I am on the subject of combining light, sound and tactile stimulation with motion, I would like to mention another interesting tidbit of information. On four occasions clients who experience motion sickness utilized the Photosonix Nova Pro combined with the IMS bed. When the motion of the IMS was moving clockwise during the session, these individuals experienced nausea and a small degree of discomfort, but when the motion was in a counterclockwise direction absolutely no nausea or discomfort was experienced. In fact, two of the individuals did not even realize the IMS was moving. Why? I haven't a clue. Could the opposite be true if tried in the southern hemisphere? Hopefully this is one of those little things that could lead to better things if enough people try this and share their experience. You just never know.

Biofeedback is the perfect modality to illustrate the effectiveness of light and sound. Let's take a moment to define biofeedback. Think of biofeedback (or neurofeedback) as a consciousness self-regulation technique (Foster 1990). This is where participants must generate the targeted brainwave at will, with no outside stimulation. Think of light and sound as biofeed-in or as a consciousness management technique (Foster 1990). This is where the mind is gently guided to the targeted brainwave state with no conscious effort on the part of the participant. With neurofeedback, the brainwave productivity of the participant can be viewed on a computer monitor, enabling the technician to observe which brain waves are being produced and at what intensity.

When an individual is doing neurofeedback training, the technician keeps track of the individual's progress. This training is broken down into seven brainwave categories: high beta (18+ Hz.), beta (15-18 Hz.), SMR (12-15 Hz.), 14 Hz. continuous, alpha (8-12 Hz.), theta (4-7 Hz.) and delta (.5-3 Hz.). When light and sound is incorporated into the neurofeedback training, certain things have been observed.

First, a direct correlation between light stimulation and the ability to produce the desired brainwave became noticeable. When the flashing lights of the light and sound instrument were in sync, that is, both the left and right LED's of the eye wear were flashing at the same time, production of the desired brainwave frequency increased. But when the flashing lights were alternating (one on/one off), production of the desired brainwave frequency decreased. This is a good tidbit of information for you to remember when you prepare your own light and sound sessions.

Second, there is a definite increase in the ability to produce the desired brainwave when the participant uses light and sound during feedback training compared to when no light and sound is used. A number of individuals were observed during their feedback training. Many showed slight improvement in their ability to produce the targeted brainwave frequency. A few displayed major increases. But the fact that every individual observed did improve on their ability to produce the targeted brainwave frequency is what matters.

This group, of which over three hundred neurofeedback sessions were performed, were unanimous in their opinion that light and sound helped their training, although light and sound was used in less than one third of the sessions. They appreciated light and sound's ability to help them achieve the targeted brainwave state more quickly, increasing the quality of the training session.

Jerry Cunningham, Ph.D., is a psychologist who utilizes neurofeedback. Dr. Cunningham has combined light and sound with neurofeedback to help entrain brain waves toward the goals being used in the neurofeedback procedures. He has found this combination to have a synergistic effect, assisting the patient in reaching the neurofeedback training goals. And as an adjunctive element to the main treatment goals, the patient has been able to take a light and sound instrument home. For example, patients who have a sleep dysfunction as part of their symptom pattern can listen to a delta session at bedtime to help them go to sleep and remain asleep. Dr. Cunningham has created a very effective method of combining light and sound with neurofeedback training, maximizing client response both during training sessions and at home.

In the field of education, light and sound/sound only instruments are used to create a positive learning atmosphere. The specific goals include: promoting the ability to learn through relaxation, enhancing motivation and visualization skills, and increasing the ability to maintain focus.

Light and sound, whether for retaining new information or overcoming test anxiety, can be a beneficial educational tool. But it can do more. It can show you how to improve your focus just by experiencing a session on a routine basis (routine meaning using the unit two or three times per week minimum). Think of using light and sound on a routine basis as a form of brain exercise. You should become aware of increased mental clarity and efficiency with less mental fuzz. Outside distractions and brain chatter should decrease. And expect mental confusion to be replaced with a sense of having everything "mentally" organized.

The only problem that I have run into with light and sound and schools are the administrators. They are very hesitant to try new technologies, which for various reasons is certainly understandable. Teachers need new technology that is successful at increasing the academic performance of students. We have proven that light and sound can satisfy this need. But if the administrators say no, then both teachers and students suffer. Because of this problem, we began utilizing sound only stimulation.

This is a safe, cost-effective way to achieve specific brain states. The instrument we used for generating sound frequencies was the Photosonix 515.Teachers find that conducting their classroom activities while simultaneously playing the 515 at very low volume enables their students to be entrained in a more relaxed state of awareness. As various research studies have proven, students in a relaxed state are capable of assimilating and retaining more information than students who are in excited or anxious states.

There are two ways teachers and students use the 515. For personal use, where the listener will be stationary, headphones are the preferred choice. If the listener will be moving around or there are several individuals listening simultaneously, then stereo (or amplified or computer) speakers work best. Sessions two, three and four of the 515 were designed to enhance academic performance. And the audio frequencies generated need not be played loud to be effective. The volume of the 515 should be about the same as the “hum” of a flourescent light.

A few weeks back I received a letter from Richard Hunsaker that pretty much describes what we all experience from listening to the Photosonix 515 while working. Richard has realized improved concentration with increased memory retention. After the first week of use, he could actually “visualize” personal improvement. Richard wrote that these improvements could not have come at a better time because he not only is the engineering manager in charge of software products for an international gaming manufacturer, but is the proud new father of a very beautiful and energetic baby girl.

Richard feels that one of the stigmas related to “self-help” techniques and products is that one must realize a problem to affect a cure or a positive result. He continued by stating the 515 improves upon good situations- making good things better. The result being enhanced and improved activities at all levels.

Light and sound can also make good things better. I have seen it happen time after time. Perhaps the best way to illustrate my point is to let you read what one client wrote about his experience. The client is Terry Keenan. Terry is a deputy sheriff. This is what he wrote about using light and sound:

“I don’t know how, but it works. This comment was made by a sheriff’s sergeant who had been injured on duty. His right shoulder caused almost constant pain. I loaned him a small light and sound instrument and a tape on pain reduction to help ease the torture.

I have found this technology very useful. As a speaker and consultant on court security I often travel. I have never been able to sleep while traveling. Nor could I catch up when I arrived at my hotel room. Because of this, I often arrived to conduct classes and security assessments in a less than alert state. Groggy from lack of sleep and attempting to substitute caffeine for energy, I'm sure I sometimes made less than an exemplary presentation.

One day I stumbled across an ad in a magazine for light and sound instruments. Wondering if they could be of any use, I went to Michael Landgraf's office. He was kind enough to allow me to try the unit there. I was impressed. After only twenty minutes on the instrument, with some soft, non-invasive music playing, I found myself awake, alert and relaxed. I bought the instrument and took it home.

My wife (Connie) and I started experimenting. We found that a twenty minute session with the instrument in the early afternoon charged our batteries for the evening. When traveling, by setting my unit to induce sleep, my travel insomnia was solved. I would settle back for several hours of uninterrupted slumber, awakening in the morning refreshed and ready to go.

My wife swears that using the machine for a couple of weeks prior to a Las Vegas vacation increased her winnings. We are going to try that experiment again!

One deputy told me that using the instrument helped him to learn to be less self-conscious in front of audiences. Another said it gave him the greatest experience of relaxation he had ever had. I would love to see the day arrive when every police building had a small, private room specifically dedicated to the use of this technology. Officers could use it to relax, help with stress, gain energy for physical exercise, study for promotional exams and focus for duty.

This is a technology whose surface has only been scratched. If given a chance, I am convinced that everyone could benefit from it. I would not be surprised if, in the near future, it gained a vast following for its abilities to reduce stress, enhance sleep (and dreaming), ease pain and so much more.

My sergeant buddy is permanently disabled. He still uses his machine if his shoulder pain becomes too uncomfortable. And he still says I don't know how, but it works."

The best way to know if light and sound technology works is for you to personally try it. Everyone is different, no two people think exactly alike. The effects of a session may be different for you compared to what you had expected.

For example, some participants were able to remember new information easier while in an alpha (or relaxed) state rather than being in the theta (or learning) state. Others have fallen asleep faster while listening to theta frequencies instead of delta, which is associated with sleeping. One individual that uses the 515 realized (or I should say his wife realized) that when sleeping while delta frequencies were generated, he did not snore, but did snore when delta frequencies were not being generated. So you never know. Experiences do vary.

You have just read how this technology can benefit people in a variety of ways. Light and sound/sound only instrumentation accomplishes this by the manipulation of brainwaves; selecting a specific session containing the targeted frequency and being gently guided to that frequency with absolutely no conscious effort.

Now you will learn how others utilize these instruments. This includes personal, business, academic and medical applications. Let these observations be your guide as you begin using light and sound or sound only. You will probably find that many of the applications described are for the same reasons you are interested in this technology.

B. Peak Performing In Beta.

Utilizing beta sessions can add to the quality of your day. You can increase your performance, your productivity, your ability to stay on task and fulfill your responsibilities. Beta light and sound sessions are energizing. They help you get going. Chemical stimulants cannot compare. Chemical stimulants have been known to produce negative side effects (i.e. physical shakiness and grumpy attitude as the effects subside). Beta/energize sessions have no negative side effects.

What about those days when your energy level is at low tide? Oftentimes pounding down the coffee doesn't work. You realize you are still at low tide, only shaky and very talkative. Your underarm deodorant no longer is effective. Your body is ready for action but your mind isn't. Try a beta session for energy. Both your mind and body will be at "high tide". We have utilized these sessions in many situations and have found them very effective for increased alertness and mental clarity. Going out to shoot some hoops or play a round of golf? Try a beta session first. The effects of the frequencies generated do not end when the session is over. They stay with you for a while. Take advantage of them.

Do you have a hard time waking up from sleep? Ever stay out too late and feel very regretful in the morning? Have you ever pulled an all-nighter cramming for an exam? These sessions are the perfect solution for those situations.

Beta light and sound/sound only sessions generate energizing frequencies that gradually increase and decrease through the beta and gamma brainwave states. Athletes and weekend warriors play these sessions before competition. They are for "psyching up" and getting motivated. Public speakers have beta sessions running while delivering presentations to keep their audiences alert and focused on what is being said. Cross-country truckers listen while they are moving their cargo across great distances to keep them mentally alert, negating the need for chemical substances. They are also used for enhancing the effects of fast-paced music. Beta sessions even help in getting the cob-webs out of your mind, allowing for clearer concentration.

Beta is the brainwave state (13-34Hz) that we associate with thinking, communicating and having an awareness of our surroundings. In our normal waking state, we are usually in low beta. There are many reasons for having light and sound beta sessions. Here are just a few:

-Use in the morning instead of coffee.
-For an afternoon energy booster.
-For depression.
-To rid your mind of mental fuzz.
-Increasing accuracy of judgement.
-14Hz/regulatory effects for your body.
-Use for motivation, to "psyche-up".
-To prepare for exercise.
-Increasing mental alertness.
-Enhancing the effects of fast-paced music.
-As an adjunct to biofeedback and neurofeedback training and therapy.

The highest frequency patterns (35Hz and up) are known as gamma brain waves. These are the brain waves that have been associated with states of peak performance. Check the Program Listing in the owners manual of your light and sound unit. Many sessions combine gamma frequencies into their design. Besides energize sessions, you will find them in creativity and visualization sessions, fun and entertainment sessions, even specialized sessions designed for overcoming depression, PMS and the occasional bouts of the "blues".

If you can locate an IQ-9110 by InnerQuest, there is a "Carnival Ride" session that is well-worth experiencing. The Photosonix Nova Pro unit has two very entertaining sessions that incorporate gamma frequencies: Photo Coaster and Mr. O's Wild Ride. Adding vibrations (tactile stimulation) to these light and sound sessions will enable you to reach new heights in mind exploration.

C. Why Achieve Alpha?

Enabling individuals to relax, yet remain mentally alert and focused are the primary reasons why light and sound and sound only instruments have been successful in my learning/relaxation programs.

We will first explore how these instruments are utilized in a medical environment, from the moment you walk through the door to adjunctive treatment at home. Then you will learn how they benefit teachers and students in both the classroom and for at-home study. Next, you will be shown three ways alpha sessions can be utilized in a business environment and finally, for personal use. As you will see, alpha brainwave frequencies are the most popular and offer a broad range of applications.

Most of the health care related programs we sponsor utilize the Photosonix Nova Pro (light and sound) and 515 (sound only) instruments. Although both are not medical tools, they have proven successful in enhancing the methods of and patient response to professional medical treatment. Scientific evidence has shown that therapy can be maximized when patients are in a less stressful brainwave state, thereby accelerating patient response to the treatment utilized.

Going to see your doctor or dentist can be very stressful. For some, just thinking about going can be difficult. You fight traffic to make your scheduled appointment . And then you wait. Isn't it interesting how many things come to mind that you could be doing other than sitting in a small room? Brain chatter. All these thoughts can have an effect on your blood pressure. Do you believe that the nurse who takes your blood pressure will receive an accurate reading? And with all those thoughts occurring, once you do get to see your physician, will you be able to completely focus on why you made the appointment?

Placing a 515 system in the corner of a reception room sounded like a good idea. By attaching two small computer speakers to the 515 and running session two (alpha/relaxation) continuously, we believed the audio tones (which were played at the same volume as the "hum" of a flourescent light) would gently guide listeners into a more relaxed state. We wanted to see if this approach would help clients experience less brain chatter and remain focused for their scheduled appointment.

Marian Thomas, Executive Director of the Los Angeles Parenting Institute, and her husband Bob, who amongst many other responsibilities finds time to lead the Institute's light and sound program, agreed to try the 515 in their reception area (which is approximately 10'X20'). The Parenting Institute is a private, non-profit, community based agency that provides counseling and education services in the areas of child abuse, child custody and family court.

We agreed that if the 515 worked in their reception room, it would probably work in any reception room. When you walk into the Parenting Institute, you will see the results of what has been happening throughout our inner-cities. Many have difficulty paying for services, many belong to gangs and many more are the victims of abuse and neglect. Obviously, there is a lot of energy in their reception room.

Over the course of several weeks, Bob observed the actions of clients with and without the 515 running. What Bob saw when the 515 was not on was a lot of movement, a lot of loud talking and general discomfort. When the 515 was running, the room was noticeably more quiet and clients were sitting around comfortably, with some clients reading while they waited (which Bob remarked as being rather unusual). What really caused Bob to realize the 515's potential was the result of the huge stuffed dog they have in that room. When the 515 was off, that stuffed dog was thrashed. When the 515 was on, that same stuffed dog was cuddled.

Promoting relaxation in clients is what the director of a successful drug and alcohol treatment program in Kansas City observed. This program is a treatment alternative to the traditional twelve step approach. The director found the 515 to be an effective background substitute for the typical muzak approach, making the clients feel at ease so they can get beyond the symptoms and move to the causative root incidents of their addictions.

Helping abused and neglected children, and individuals overcoming drug and alcohol addictions relax while waiting for their appointments are strong indicators that a sound only instrument can be a useful tool for creating the desired reception room environment. Both situations illustrate how the generation of alpha frequencies can effect individuals in a positive manner.

How do you use light and sound/sound only technology while you are treating a client? It really depends upon why the client has scheduled the appointment. Oftentimes a ten to fifteen minute alpha session immediately before therapy will allow the client to become focused. When major stress is apparent, a theta session would be more beneficial.

If you are a chiropractor, allow your client to experience a ten to fifteen minute alpha session to relax, yet remain mentally alert, before beginning treatment. Acupuncturists can utilize alpha sessions much the same way. Physicians, counselors and therapists will also notice heightened patient response to treatment when utilizing this approach. Having a sound only system running during group or individual counseling sessions has proven to be very effective for staying on the topic being discussed. And here is something to consider: when you visit your dentist, wouldn't it be nice to experience a relaxation session combined with music to occupy your thoughts during treatment? For many individuals, time seems to go by faster when using light and sound during dental work, making the whole process easier to endure.

Think about it. You work all day, rush to make your scheduled appointment and have several things remaining on your "To Do" list. Can you remember everything you intend to ask your therapist? If having a sound only unit in their reception area or access to a light and sound unit immediately before treatment was available, would you take advantage of it? Being relaxed, yet mentally alert with increased focus on the task at hand would certainly benefit both the client and the medical professional. The next time you schedule an appointment, mention this technology. Share what you have learned. Oftentimes medical professionals are not aware of the benefits from utilizing these instruments.

Jocelyn Roland, Ph.D., is a clinical psychologist in the psychological services division of a major Southern California law enforcement agency. Dr. Roland utilizes light and sound during treatment primarily for general stress reduction. She also uses light and sound to reduce anxiety and phobic disorders through adjunctive use of this technology to therapy and systematic desensitization. In still other endeavors, Dr. Roland uses light and sound for sleep disorders, learning and memory, and general relaxation.

In 1994, Dr. Roland, Sandra Levine and I worked together to find new ways of utilizing light and sound for the benefit of police officers. In a three month study, we invited personnel from the communications division (radio and emergency radio and telephone operators) of a local law enforcement agency to participate. Thirty-five 911 Emergency Operators allowed us to measure their "base-rates" before and after experiencing light and sound sessions. These base rates measured their heart rate, body temperature, left and right brain states and left and right galvanic skin responses. Because these operators are under tremendous pressure to think fast and act quickly, our goal in this study was to see if using light and sound during breaks would enhance their relaxation and coping skills while improving their focus. We also included a variety of audio cassette tapes that dealt with meditation, visualization and guided imagery techniques that could be played in conjunction with the light and sound sessions. These 911 Operators found listening to relaxation tapes while experiencing light and sound sessions so beneficial that the 911 Trainees began receiving light and sound orientation while still attending classroom instruction. The program itself was rated excellent by the participants, with everyone experiencing greater control over stress than before.

In 1995, one clinical psychologist submitted observations concerning her use of light and sound instruments as an adjunctive tool in her counseling with police officers. As you probably realize, this psychologist's client population is a skeptical one. It took real results for them to accept light and sound. The psychologist utilized light and sound in her treatment of officers suffering from overwork, burn-out, post-traumatic stress, shift changes (interrupting sleep patterns) and officers studying for promotional exams.

The mixture of alpha and theta light and sound sessions proved effective and beneficial for these officers. She also found these sessions extremely useful for clients suffering from chronic pain and post-traumatic stress disorders, observing that it has helped them to relax, to focus on tasks and to sleep.

It is apparent that the prime purpose of using light and sound or sound only before and during medical treatment is for helping the client to relax and focus on the treatment (or in the case of dental work, helping the client to think of anything but the treatment). Alpha, and sometimes theta, are the frequencies utilized most. The 515, in regards to the programs we are involved with, is the instrument of choice. The reasons being: it is far easier to set up and run compared to light and sound, it is less expensive than light and sound, and it can be safely used by anyone. In essence, it saves time, money and alleviates the possibility of someone experiencing any adverse effects.

Now that we have covered how this technology benefits the client from the time they first walk through the office door to the time they have finished their treatment, let's find out how light and sound/sound only can be of benefit to the client when they step outside that office door.

When a client is handed a light and sound unit by their therapist for use at home, the problem has always been the variety of built-in sessions contained in the units. Say you have a unit with twenty built-in sessions. How many actually pertain to the reasons for letting them use it and how many sessions do not? Do you think telling them not to use sessions one through three, for example, is going to keep them from listening to those sessions first? Faith alone just doesn't make it.

For the last ten years I have listened as the manufacturers of light and sound questioned why their instruments were having such a hard time being accepted by the medical community in general. But how can these professionals apply this technology if they are unable to control which sessions their client chooses?

Finally, one of the manufacturers has addressed this need. The instrument is called the Pro Tutor. It is an "empty" light and sound unit. Same capabilities, same everything, except it does not contain any built-in sessions. The therapist can either create particular sessions to meet the specific needs of their client or choose sessions already designed and listed on several computer disks.. These sessions are then displayed on a computer menu that can be easily downloaded right into the Pro Tutor light and sound unit. It is the only model available that makes guaranteeing what sessions clients listen to possible to control. It's about time.

So in the past, yes, we have worked on faith alone. And yes, oftentimes the clients did listen to sessions that were not recommended. Being individuals who are naturally curious, I could not fault them, only redirect them into focusing on why they are doing light and sound sessions in the first place.

The residents of Our House, AIDS Project/Los Angeles' in-house treatment facility, are good examples. All residents suffer from at least one addiction, plus they are HIV+ or have AIDS. InnerQuest light and sound instruments were always available to them, anytime they felt the need. And these guys tried all the sessions at one time or another. They were very curious about what kind of effects they would experience with the different sessions, and were open to discussing the sensations they received. The consensus during these discussions was that the alpha and theta sessions gave them the most relief and were the most satisfying.

When describing the experience of light and sound instrumentation, Gerramy Quarto, former Treatment Coordinator for Our House said it best: "Residents of the program here at Our House have described achieving a state of relaxation that they have not experienced for quite a long time. As you realize, with our six month residential HIV/Chemical Dependency Program, overcoming stress and learning to relax is the primary step towards positive personal development. I believe these units to be a tremendous help to the residents."

What all this means is whether you are listening to sound sessions while you are waiting to see your physician, doing light and sound as part of your treatment or doing sessions at home to augment your treatment, alpha sessions appear to work best at relaxing and focusing a client. As a result, the patient's response to medical treatment is enhanced and accelerated.

You have undoubtedly heard how some light and sound sessions are being combined with other technologies to help those with ADD, ADHD, migraine headaches, chronic pain and a myriad of other disorders. Researchers are developing new protocols utilizing light and sound sessions to help individuals who suffer from these disorders. And this is great news. In time, it is reasonable to expect more people will be able to use this technology rather than being prescribed pills, which have been known to create negative side effects in many situations and is not as cost effective as light and sound. If you are interested in learning more about how these researchers are using light and sound to develop new protocols, please refer to the Suggested Reading section.

Let us now examine the ways alpha light and sound/sound only sessions can enhance academic performance. Test anxiety can be a real pain when you are trying to learn new information. Brain chatter, mental fuzz and fatigue can all factor in destroying any gains you make while preparing. What can you do? A simple relaxation (alpha) session lasting no longer than fifteen minutes. One before and one after your study session should improve your learning abilities.

Examples of using light and sound to overcome test anxiety include Jocelyn Roland, Ph.D., and her experience when preparing to complete her doctorate in psychology and Janet Davis, when she was preparing for the California State Bilingual Teacher certification exam. Both women have excelled in academics, but both were rather stressed as they prepared for their exams. I remember talking to Jocelyn a few days before her test. She said she was ready, but the usual anxiety she had come to expect was not there. She felt almost as if she did not care, which was hardly the case. Jocelyn realized the difference was not experiencing any anxiety. She was successful in passing her examinations and has continued to utilize light and sound as a means of controlling daily stress.

Janet, in comparison, was experiencing test anxiety for the first time because she was having trouble acquiring enough fluency in Spanish to pass the state exam for bilingual certification. After incorporating light and sound into her studies, she first noticed a decrease in the physical symptoms associated with test anxiety and the accompanying mental confusion. Then Janet noticed an immediate increase in her ability to retain new information. She found that her method of studying with light and sound made retaining and retrieving test information more effective and took less effort than when not utilizing light and sound. Janet was also successful in passing her certification exam and she too, continues to utilize light and sound in her daily life.

Most teachers that I have worked with invariably take the light and sound instrument home for their personal use. Steve Brager, Janet Davis and Lyne Avignone-Young, all elementary school teachers in the L.A. Unified School District, use light and sound at home for different reasons. Steve uses light and sound to relax and increase his focus, believing this has helped him become a better teacher. Janet uses her instrument while studying new material and working at her computer. Lyne utilizes her unit to help lower her blood pressure and to relax. Incidentally, Lyne also uses her unit to help her three year old son fall asleep at bedtime.

If you are a student or are studying for a promotional exam, try utilizing alpha frequency sessions. You should find that any test anxiety you are experiencing will be decreased or negated altogether. And remember that alpha sessions can be highly beneficial before, during (with sound only or with open-eyed LED frames on your light and sound unit) and after studying.

Here is a description of how this technology (alpha sessions in particular) has been applied in various businesses that I have worked with. There are a variety of ways to incorporate light and sound and/or sound only instrumentation into a business environment. The following methods have been quite successful.

Many companies have developed stress management seminars for their employees, others have turned "break rooms" into relaxation areas and others allow personal systems into the employee workplace. Utilizing light and sound/sound only instrumentation in the work environment is a win/win situation for the employer/employee because the more relaxed from stressors a person is, the more resourceful, accurate and productive that employee becomes.

Employers who have incorporated learning/relaxation tools and/or seminars into their company have observed an increase in employee relations and productivity, and a decrease in employee absenteeism due to stress. Considering medical and research doctors estimate 70% to 90% of all illness is due to stress, these instruments become not only cost effective, but also increases the employee effectiveness within the company.

Stress management seminars for employees include a variety of tools and techniques for coping with daily stressors, improving on-task focus and achieving peak performance. Light and sound instruments can be used to attain each one of these goals. As mentioned previously, fifteen to twenty minute alpha light and sound sessions will allow you to relax, yet remain mentally alert and focused. For achieving peak performance, research indicates the utilization of many frequencies, mainly gamma. More information concerning peak performance can be obtained through the Suggested Reading list.

Stress management seminars are as varied as the personalities of the individuals who present them. As mentioned, they all offer a wide range of methods and techniques to help you cope with the idiosyncracies of your job and the effects your job has on you personally. Should you be interested in starting a seminar of this nature at your work, there are many ways to begin. Does your work offer medical coverage? If so, talk to the insurance carrier about how to create stress management seminars. Your local county mental health department should also be helpful. So too are the hundreds of books written on the subject of overcoming stress. Consulting a professional therapist is also a good starting point.

There are dozens of questions you need to be concerned with when considering seminars of this nature. These professionals will help answer many of them for you. However, please keep in mind that after seven years of personal observations, I have seen many of these seminars fail at their goals simply because workers did not utilize the methods and techniques they learned from attending the seminars. Whether they were too busy keeping up with their job responsibilities or did not take the time to incorporate the methods into their daily job routine, it really does not matter. It all boiled down to "out of sight- out of mind". Stress management seminars can be very beneficial. You are shown techniques that improve both your job performance and quality of life. Put to use what you are shown.

Everyone, at least once a day, seems to visit the break room. Does your break area have comfortable chairs or a sofa for relaxing? What color paint covers the walls? If you enter your break area and the last thing you feel is relaxed, do something about it! Most employees do care about their working conditions and would be willing to help make changes that would be beneficial for everyone. Take the initiative. A handful of employees on their day off can turn a break room into an area conducive to relaxation. And the long-term benefits of volunteering one day to do this can be remarkable. No longer do you have to sit on metal chairs and look at walls that scream "get back to work"! After all, the employee break room is for you and your fellow employees, right?

Now here is an inexpensive way to put light and sound machines into your break room. Call your local light and sound dealer and explain what you are doing, the reasons for doing it and how it can be of benefit to the dealer (if we have units at work, many will want units at home). They should work with you. If there is no dealer in your area, call light and sound manufacturers. Offer to have questionnaires for employees to fill out that you will make available to the manufacturer. This is a huge benefit to them for a number of reasons. Most important, they need feedback like this to help in their continuing efforts to improve on the sessions and user friendliness of their instruments. I have never talked to a light and sound manufacturer who wasn't willing to listen and help where they could. All of you can benefit from this approach.

Other things will also be needed. Make copies of your favorite tapes so you and others can enjoy them. Visit your local retail discount store and purchase a couple of cassette walkmans. I have seen some selling for as little as nine dollars- and they include headphones. Right now you are undoubtedly thinking sure, a few of us will take the initiative and get burned. The equipment we donate will be stolen. Okay, I too have seen that happen and it really stinks. But I have also seen the long term effects of what will happen if a small group of employees take the chance, of believing in their co-workers. Trust happens. It is like a seed you plant in spring. You cannot see it at first, but then it grows. It gets bigger and branches out. And it is beautiful. Soon everyone wants to take part in caring for it. The few co-workers who seem intent on destroying the good thing you've started will disappear or have a change of heart. Your employer will take notice. Good things can happen. All you need is to have a little trust, a little patience and the strength to get things started.

Oftentimes an employer cannot afford a professional to conduct stress management seminars or relaxation areas are not possible because of space. For these situations, having access to light and sound technology at your workstation can prove to be extremely beneficial. Use the Photosonix 515. Connect this hand-sized instrument to two small speakers and listen to session B2 (continuous alpha frequencies). If part of your workstation includes a computer, the 515 system has a stereo patch cord that connects directly into the audio input jack of your computer terminal. The 515 can be played on any computer that has a sound card and will not affect any computer program that you are running. And remember, when doing a 515 session, the volume is kept low. No one will be inconvenienced by your using this instrument. Since 1995, several individuals have utilized the 515 in this manner and have told me that it has helped them maintain focus on what they were working on, has made doing their paperwork easier to complete and has kept them from "totally losing it" when problems arose.

In June of 1995, a clinical psychologist working in the psychological services division of a major Southern California law enforcement agency was using the 515 every day to relax. She had it running in her office when she was alone or with particular clients. All of that changed when she let her front office staff use the unit on a particularly stressful day. They now keep it up front with them, and have it going most of the day for relaxation purposes.

In 1996 we used the 515 in a major furniture and sofa factory showroom. Not in the area where the customers shop, but where the employees do their paperwork. Session B2 was run throughout the workday. Over the course of several weeks, the store manager made several interesting observations. There was less bickering amongst the employees. Paperwork was completed in a timely manner with less mistakes. The day felt like it went by faster. They made more sales. The 515 is still being used by that company.

Is there a lot of movement and noise in your workplace? Could the 515 be effective in such an environment? The hair salon I frequent is a good example. The hair stylists are constantly moving and the conversations can be quite loud. So a 515 was placed in the corner and session B2 (continuous alpha) was run during working hours. The major difference observed was the volume. People were still talking about what was happening in their lives, but they were not in competition with each other to be heard.

A good illustration of the difference between using and not using the 515 in this salon was something the employer shared with me. She knows her employees well and can sense when one begins to get anxious. Thursday evenings are one of their busiest times and one stylist had appointments scheduled from four in the afternoon straight through to eight-thirty that evening. Her appointments began to get backed up and her clients found themselves having to wait. Normally, the employer can see the panic begin to develop, observing a sense of urgency in the stylist's mannerisms. But this time was different. There was no panic. Instead, the stylist took her time making sure each client received exactly what they wanted, just like when her appointments are not spaced so close together. And even though her clients did have to wait, no one became upset or asked to be rescheduled. Everyone completely understood. This employer believes the difference in attitude was the result of using the 515.

It is apparent that utilizing alpha frequencies while working or during workday breaks increases one's ability to cope with the responsibilities of one's job. This technology, whether using light and sound or sound only instrumentation, increases our ability to focus on the task at hand, remain relaxed yet mentally alert, and not allow outside distractions to interrupt our daily job responsibilities.

Regardless of what you do to earn a living, why is it that when you do something really good you are rarely "patted on the back"? But when you mess up, what happens? Yes, we hear about those every time. How do you handle that? When you arrive home, do you pour yourself a stiff one? Kick your dog? Is that how you really want to deal with your stress? Here is an alternative. Have your 515 connected to computer speakers. When you arrive home, turn the unit on and press session A2 (for relaxation).

Now go hug your dog, then kick back and relax. Let all those stressors from your day just wash away. You need to take the upper hand at controlling your stress. This is a good way to begin. Make this part of your daily ritual when you get home rather than doing quick fixes. You will live longer (and so will your dog).

What is your drive to and from work like? If it is a breeze, congratulations! Consider yourself fortunate. However, if you live in a large city and have to deal with stop and go traffic, the following example just may help you.

Gordon Smith is a longtime friend who for years has lived about twenty miles outside Las Vegas. Gordon is a business owner who has seen this city grow from a small gambling town to a major urban area. Anyone who has been to Las Vegas has personally experienced the bumper-to-bumper traffic and traffic signals that take forever to change. But Gordon takes it all in stride. He has his 515 in the seat next to him and runs the alpha session (A2) during his commute. He accomplishes this by first charging his 515 batteries at home (the charge lasts a really long time) and then connects the unit to computer speakers. He never plays it loud and he never listens to any other session while driving. Why? No need. Alpha puts him in the right frame of mind to tolerate "road warriors" and prepare for the work that awaits him. And during Gordon's commute home, the 515 assists in helping him wind down.

I refer to sessions utilizing alpha frequencies as the universal brainwave sessions because they are used for such a wide range of applications and everyone I know that has a light and sound/sound only instrument tells me they listen to these the most. We have used alpha frequencies in retail stores, business reception and work areas and during both individual and group therapy and counseling. They are used for learning, for relaxing and for meditation. People use these sessions while at their computers. They are used to enhance "mood" music, for overcoming test anxiety and as an aid in lowering blood pressure (even an assistant coach in the NBA utilized one of my Tutor+'s to help control his pressure). But most importantly, they were designed for general stress reduction and they have proven extremely successful in doing so.

For general relaxation do sessions that target frequencies in the 4-12Hz range (theta and alpha states). What frequency session you choose really depends on how you are feeling just before you begin the session. Think of it this way: the lower the frequencies, the deeper the relaxation experience. If you really need to relax, do a theta (4-7Hz) session. If you would like to relax, do an alpha (8-12Hz) session. And if you are not sure, do a session that combines both.

Here are some situations where doing an 8-12Hz (alpha) session beforehand could help:

-Before taking a test.
-Making a presentation.
-Meeting your ex-spouse.
-An important appointment.
-Making a commitment.
-Housework/cooking dinner.
-Going to see your dentist/physician.
-Driving.
-Relatives are coming over.
-Anything needing increased focus or concentration.

Here are some situations where doing an 8-12Hz (alpha) session afterwards could help:

-Coming home from work.
-Exercising/working out.
-Supervising children.
-You just saw an accident.
-Balanced your checkbook.
-Experienced an earthquake.
-Studying/learning new information.
-Experiencing muscle tension.
-Weighed yourself.
-Had too much caffeine.

D. Discoveries Await You In Theta.

One of the most effective methods for studying or preparing for an exam is to use an alpha or theta session with an accompanying audio cassette. This audio cassette is taped by you, with the notes and information for your test spoken by you. Many times the mind will tune out a voice for any number of reasons, but research suggests that by recording the information in your own voice, your mind will listen.

This is a good study technique because you first review your notes before recording, then review them during recording, plus how many other times you listen to the tape while you run a light and sound session. Try a theta session before trying alpha while playing your cassette as theta is the "learning" state. However, keep in mind that we are all different, so if theta does not feel right, then try alpha. Your goal for using light and sound with your personal cassette should be the assimilation of new information with high memory retention. Do not overdo these study sessions. Try to take a ten to fifteen minute break for every hour you study.

Steve Brager is a certified Resource Specialist and Special Education Teacher in the Los Angeles Unified School District. In 1991, we began a program utilizing light and sound for his hyperactive and learning disabled students.

Many of Steve's students are bused from outlying areas and still others are from one-parent homes with little or no supervision. Oftentimes they are not prepared for school, which makes Steve's responsibility of teaching them that much more difficult. Light and sound has helped. We have used, over the course of this program, such light and sound instruments as the IQ-3 by InnerQuest, the Esprit by Synetic Systems and the PR2-X by MindGear. All three of these units have quality learning sessions.. We have also used the Photosonix 515 sound only instrument for creating the desired classroom atmosphere.

What we have observed from Steve's students as they use light and sound is the hyperactive students have been able to complete assignments without losing focus, and they are able to stay on task for longer periods and focus their concentration on a given skill. These students were bound by their disability in the past and their self-esteem suffered. Some students are now eager to learn as they are beginning to recognize their capabilities. Many have learned to reduce their stress and anxieties, resulting in a mind set more conducive to learning. Reducing stress and practicing relaxation as a skill has also been a vehicle to help students transcend problems at home that cause road blocks to learning.

One student in particular, who was having a hard time controlling his temper at home, came to me one day and with a very big smile described how light and sound enabled him to realize when his anger at others would begin to develop. Unfortunately, his anger was usually directed at his grandmother, because she was there when he would arrive home each day from school. That day when he arrived home, instead of getting mad at his grandmother, he went up to his bedroom and laid down for a few minutes. As he was lying there, he closed his eyes and remembered how he felt when he was doing a light and sound session. This enabled him to calm down and relax. Then he went back downstairs and hugged his grandmother.

Both Steve and I have been pleasantly surprised by how many parents have come up to Steve and said "We don't know exactly what you are doing with our children, but their behavior at home has gotten better. Please keep it up." Remarks like that make sponsoring light and sound programs worth it.

So whether you own a light and sound unit or a sound only system, utilize it's theta sessions while you study. The benefits include: improved focus on the task at hand, increased memory retention and a reduction in anxieties associated with taking tests.

For personal relaxation and mind exploration, many light and sound users enjoy the sessions listed under Meditation. Especially the sessions that feature the Schumann Resonance Frequency (7.83Hz). This frequency is generated from the energy of the earth as it resonates with our upper atmosphere. No one really knows for sure why this frequency is so powerful. Your Program Listing will have it included in such sessions as Earth Harmony, Earth Grounding, Chakra Balancing or even under Schumann Resonance, believe it or not. Regardless, there is a very special way to experience this session that should have a dynamic impact on you.

First, make sure your unit is fully charged. Then take it outside and find a nice comfortable spot on the ground. Lay down with your head to the north and your feet to the south. Then enjoy the session.

Incidentally, if you enjoy watching the X-Files, you may remember an episode they did about an individual who acquired the ability to control lightning (#52/"D.P.O"). In it, they mentioned the frequency of lightning was 7.83Hz, the Schumann Resonance Frequency.

The women at Recovery Serenity Ranch would do the Schumann Resonance session outdoors during the spring and summer months and found it extremely refreshing. Although the women that stayed at RSR were the victims of abuse and addiction, from the beginning they utilized light and sound with great expertise, oftentimes teaching the teacher.

Each resident was able to use these instruments on a daily basis and attended a weekly seminar devoted to maximizing the benefits. This approach allowed each individual to proceed at their own pace. Topics discussed in the seminars mainly involved using light and sound technology correctly, experiencing sessions aimed at recovery and learning a variety of relaxation techniques for use when they graduated. The purpose of this program was to give these women a built-in relaxation response to override any anxious or stressful situations that may occur after they had left RSR.. In addition, this program enabled them to experience the benefit of complete mind/body harmony.

Theta sessions are often referred to as the learning sessions. However, they too are used by musicians when composing music, writers seeking new ideas and painters in need of inspiration. Meditation can be more profound. Soft rock and jazz will be enhanced. Listen while you are playing a video game (talk about being focused). Play theta sessions with your self-improvement tapes. In any of these situations, your experience should be heightened.

Here are examples of when 4-7Hz (theta) sessions can be beneficial:

-Experiencing headaches/pain.
-For problem solving.
-Enhancing visualization skills.
-After working at your computer.
-Your teenager knows everything.
-Seeking behavior modification.
-Experiencing forgetfulness.
-Nervous.
-Business readjustment.

-Composing music or painting.
-Had a very stressful day.
-For deep meditation and relaxation.
-After driving many hours.
-Feeling imbalanced.
-Activating personal goals.
-Irritability.
-Facing major life change.
-Increased responsibilities.

-Deep mind exploration. -For personal healing.

These are actual ways that clients and friends have utilized theta frequencies. It is by no means a complete list, but should offer you ideas on how you can personally benefit from this brain state.

You should also utilize frequencies in both the alpha and theta brainwave states. These frequencies gently guide you back and forth, from one state to the next, in an effort to inspire your creativity. Again, these sessions are used by writers, musicians and painters. Students and teachers enhance their academic abilities. Meditative experiences become maximized. Visualization skills are enhanced. Reading for enjoyment is at it's best. These sessions are used to create the ideal learning environment. When searching for new ideas, use this combination of alpha and theta frequencies.

E. The Significance Of Delta.

What happens when you are ready to go to sleep? Do the activities of your day repeat themselves in your mind? Do you think about what needs to be done tomorrow? Do you find all this thinking is keeping you from going to sleep and you end up getting more anxious? Try a sleep session. Stop trying to focus on things and just let your mind drift. So many people have benefitted from this approach.

Delta frequencies are for going to sleep. If you suffer from any type of sleep disorder, such as insomnia, bruxism (teeth grinding) or experience nightmares, seek professional medical help. Your light and sound unit can be the ideal "at home" adjunct to the professional treatment you receive. Dr. Jerry Cunningham has observed that utilizing light and sound in this fashion will assist patients who have a sleep dysfunction as part of their symptom pattern get to sleep easier and remain asleep.

Medication that aids us in falling asleep can often linger through the next morning. Delta sessions do not do that- you awake refreshed. Try these sessions for jet lag or for shift changes. These delta frequencies are all you need when you cannot fall asleep on your own.

Delta brain waves are the deepest frequencies (.5-3Hz) we produce. When we are in the delta brainwave state, we are asleep. All kinds of external noise can prevent us from going to sleep and so can internal "brain chatter". How do you deal with it? Do you take a pill? They have certain disadvantages. Some pills that help you fall asleep have a morning after "hangover" effect that can impair your concentration. They can be dangerous should you awaken during the night and find you must get out of bed. Sleeping pills may also induce tolerance and dependence, and they may interact dangerously with other medication and alcohol. Light and sound or sound only instruments are safe and effective alternatives for falling asleep.

Make sure the light and sound/sound only instrument you purchase has at least one session to help you fall asleep. You could live in the middle of nowhere with only the chirping of crickets in the background, but if your mind is active, going to sleep can be difficult. Here are some reasons for using a sleep session:

-Shift change at work.
-New surroundings/staying in a hotel.
-Over tired.
-Jet lag/crossed too many time zones.
-Too much outside noise/barking dogs, sirens, etc.
-Have a new family addition (infant, puppy).
-Cannot stop thinking about the day's events.
-Thinking about what will happen tomorrow.
-Ate too much/cannot get comfortable.
-Feeling excited/have nervous energy.

Since becoming a light and sound dealer, many of my clients have gotten married and started families. A newborn at home can really turn "sleep time' into turmoil. In an effort to regulate the newborn's sleep cycle, many have utilized delta frequencies. They also use it for themselves. And as their newborns have become toddlers and infants, getting their little ones to go to sleep can at times be very difficult. The sleep sessions are perfect for times like these, for both you and the infant. Do they wake you up in the middle of the night because they are thirsty? Many of us have been in that situation. If there is a problem, fine, we deal with it. But how many times has the real reason been because they like to see mommy or daddy try to function while still in the throes of sleep? Run a delta session! When using the 515, you turn the unit on when they go to bed and it generates delta frequencies until you turn it off. Several parents have tried this approach and it really works. The session is like a sleep trainer. Go to sleep. Stay asleep. Get a good night's rest. That's what it is all about.

III. HOW TO CHOOSE A LIGHT & SOUND OR SOUND ONLY INSTRUMENT

Here are the four leading manufacturers of light and sound/sound only instruments. They are here to assist you with any questions that you may have regarding this technology:

Comptronic Devices Limited: 403/450-3729
Microfirm, Inc.: 562/865-8582
MindGear, Inc.: 216/354-5159
Synetic Systems, Inc.: 206/632-1744

A. What Type Of System Do You Need?

Now that you understand what a light and sound instrument is, you must ask yourself what you need it for.

If you need to relax, you do not need an expensive machine. There are many instruments available that will meet your needs. These units cost between $150 and $200. Most of these instruments contain sessions in the beta, alpha, theta and delta brain states. Most are made of hard plastic while others are of metal. Both are durable, have very few moving parts and should last for years. These instruments are simple to operate and are perfect for beginners or for those of us looking only for basic sessions.

Often times I will take one of these "beginners" units with me on camping trips. Not that it is needed in the great outdoors. Sometimes it is nice to combine the outdoors with a light and sound experience. Especially after a session. Anyway, I dropped a unit in a river once. It was soaked, but after drying out for a day and a half, it worked just like new.

So trust me when I say these might be inexpensive units, but they are very dependable!

If you are interested in mind exploration, then get an instrument that offers a variety of built-in sessions. Most have space available for you to custom design your own sessions. Probably the greatest advantage to having a more advanced model is the fact that most all of them have computer interfaces. This allows you to view a series of sessions that are contained on computer disks via your monitor, select which sessions that will meet your present needs, then download them directly to your light and sound instrument. You will have the opportunity of increasing your session library without the need for purchasing a new instrument. Light and sound units that fall into this category have a wide range of price. Many sell for around $300. Others can cost more than $700. So educate yourself. Shop around. They are all well worth their price, just make sure you are happy with your choice.

There are intermediate models available. These too have their advantages. Some are for specific needs or are for specific purposes, like the Pro Tutor. It all depends on what you want light and sound sessions for and how much you are willing to spend.

I do not believe there are that many quality sound only instruments available. By quality I mean possessing the ability to keep the mind entrained for specific periods of time. There are two instruments that are worth the investment. The Photosonix 515 and the Audio Imagery. These instruments are great for using with either headphones or speakers, enabling listeners to walk around while enjoying sessions instead of having to remain stationary (as when doing light and sound sessions) and for creating the desired "atmosphere" in a room or fairly large area. They are also beneficial to those who are light sensitive or are unable to use light and sound because of possible adverse effects.

We are aware of other sound only instruments that are available. There are some inexpensive models out there. There are some expensive models out there. We believe the best way to tell if they are effective is to use them while the listener is hooked up to an EEG/neurofeedback instrument.

That way, the brainwave production of the listener can be viewed on a computer monitor. We have done that. And regardless of their price, others are not as effective as the 515. They do not possess the capacity to generate dual binaural beats in an independent fashion. The 515 does. We have seen the differences via EEG/neurofeedback, and other units fail to keep the mind entrained because they lack the ability to produce the necessary frequency variations that keep the mind from getting "bored". Whereas the dynamics behind dual independent binaural beats allows enough frequency variation to keep the mind interested.

What I find most advantageous about these sound only units is how widely applicable they are for the investment. As an illustration, just compare what they offer for their price to the cost of an alpha chamber!

B. What To Look For.

Do you realize how many different colors of LEDs there are available for the eye wear of light and sound instruments? There are red (the majority of users enjoy this color the most), green (not as intense as red but still pretty good), full-spectrum (everyone should give these a try), white (going for the ganzfeld effect), clear or see-through (a definite must just for diversity) and then you have the rest (don't waste your money). Every color but blue. I have always wanted blue LEDs but have been told they cannot produce the necessary intensities, they do not last or they are impossible to mass produce. It would be interesting to experience light and sound with a kind of "black light" stimulation. As you can tell, there are plenty of choices available when selecting a light and sound system. Buy more than one type. Add diversity to your light and sound experience.

Headphones are a different story. Regardless of the manufacturer, if you buy an inexpensive system, you will receive inexpensive headphones. Because the sound quality is so important to brain entrainment, you would think the manufacturers would put more effort into including a quality set of headphones into their product. They don't. You suffer. Even purchasing an expensive system does not guarantee that you will receive a quality long-lasting set. It drives me nuts when I'm doing a session or assisting someone else and a loud crackle happens. Ruins the whole effect! So here is something to keep in mind. Keep the headphones that come with your system as a backup. Watch for sales at your local Circuit City, Radio Shack or neighborhood Audio/Video Dealer and purchase headphones that you really like. Go for comfort and how they block outside noise. If you buy wireless, you will not be happy.

The most important factor to consider when purchasing a light and sound or sound only unit is: Do you understand how to use it? If you do not know how to use your unit, all it will do is collect dust.

You must also be aware of what the instrument's capabilities are and if they meet your needs. It is so important for you to learn as much as possible before purchasing a system. Refer to the Suggested Reading section to gain more information pertaining to the areas of your interest. Remember: The better informed you are, the better choice you will make.

C. Different Types Of Stimulation:

Not all light and sound or sound only instruments have the same abilities. Most, for example, do not possess dual binaural beat capabilities. So what types of sound are generated by these instruments? How does the sound generation interact with the pulsating lights? And where does tactile stimulation enter into the picture?

1. Audio Frequency Stimulation.

We are all aware that an unborn child is capable of hearing. Our sense of hearing begins long before our sense of vision. As such, sound becomes a powerful form of stimulation for achieving desired brainwave states.

Sound stimulation can come in many different forms. Some of the most common include: surf or pulsed white noise, pulsed tone, white noise and tone combined, pulsed chord, binaural beats, dual binaural beats and white noise with either binaural beats or dual binaural beats.

Nearly all light and sound instruments are capable of generating white noise. You know that fuzzy sound your television makes when you are between channels? Or the sound rushing water makes as it flows by you? That's what pulsed white noise sounds like. Pulsed tone is just like white noise except a pitch is utilized. Your light and sound instrument should have the ability to let you adjust the pitch. Combining pulsed tone to pulsed white noise is often used in "entertainment" sessions for creating special sound effects. A pulsed chord is basically the combination of two pulsed tones that are set apart at musical intervals.

A binaural beat is the combination of two audio frequencies or tones. One tone is directed to the left ear while the other tone is directed to the right ear. Say the left ear is receiving a tone of 180Hz and the right ear is receiving a tone of 190Hz. What the brain actually "hears" is 10Hz, which is the difference between the two and is known as the offset frequency. Dual binaural beats are created simply by combining two totally independent binaural beats. The resulting four channels of audio beats work together to generate several offset frequencies. The resulting frequency variations are very dynamic and complex, and have become the ideal method for maximizing brainwave entrainment. In fact, dual binaural beats are so effective that listeners can utilize stereo speakers rather than headphones and still receive the full effect. White noise is sometimes combined with either binaural beats or dual binaural beats. This method is limited, however, to session designs targeting relaxation and meditative states.

Phase settings for pulsed tone and white noise are connected to the light settings. By combining sound with light, you can utilize eight different phase settings: in phase, out phase, cross phase, sound out phase with lights in phase, sound in phase with lights out phase, clockwise motion, counterclockwise motion and front/back phase. All phase settings are rather self-explanatory and are described in more detail in the owners manual of your light and sound instrument. Phase settings with dual binaural beats involves a little math and some experimenting. For information on creating sessions utilizing binaural and dual binaural beats, refer to the Suggested Reading section.

Left Channel On / Right Channel On

I I

Standard In-Phase

Left Channel On / Right Channel Off

I O

Standard Out-Phase

Left Light On / Right Light Off Left Sound Off / Right Sound On	Left Light Off / Right Light On Left Sound On / Right Sound Off
I O	**O I**
O I	**I O**

Cross Phase Light and Sound Stimulation

Both Light Channels On Left Sound On / Right Sound Off	Both Light Channels On Left Sound Off / Right Sound On
I I	**I I**
I O	**O I**

Sound Alternating with In-Phase Light Stimulation

Left Light On / Right Light Off Both Sound Channels On	Left Light Off / Right Light On Both Sound Channels On
I O	**O I**
I I	**I I**

Sound In-Phase with Alternating Light Stimulation

I O	**O I**	**O O**	**O O**
O O	**O O**	**O I**	**I O**

Clockwise Light & Sound Stimulation 1=On O=Off

I O   O O   O O   O I
O O   I O   O I   O O

Counterclockwise Light & Sound Stimulation   1=On   O=Off

I I   O O
O O   I I

Front / Back Light & Sound Stimulation

2. Pulsed Light Stimulation.

Light and sound instruments effect brainwave activity through the utilization of pulsating lights, pulsating audio tones and offset frequencies. These effects are achieved by using light-emitting eye frames and stereo headphones.

Light and sound manufacturers produce three types of light-emitting eye frames for generating the pulsating light stimulation. These types of eye frames are known as: standard LED (light-emitting diodes), open-eyed LED and Lumasoft.

The standard LED eye frames look a lot like sunglasses on the outside, but have LED's mounted on the inside of the shaded lense areas. There are usually four LED's mounted in the middle of each lense. These eye frames are designed for use with your eyes closed and can be utilized with any light and sound session.

The LED's mounted in these standard eye frames come in a variety of colors. Many light and sound users prefer experiencing sessions with red LED's because red produces the brightest intensity. Still others believe the red LED's help the mind's ability to generate a wider range of colors and patterns, enriching their session experiences. By far, red LED standard eye frames are the most common.

There are alternatives. For a less intense and more aesthetic experience, many prefer full-spectrum eye frames. These eye frames consist of two red and two green LED's mounted on each lense. It would be well worth your time to experience a session using full-spectrum eye frames. You should find it extremely relaxing.

Standard eye frames come with other colors of LED's. All green or all amber, for example. While these colors are unable to produce the same intensity as red LED's, they are preferable to users who experience some degree of light sensitivity.

The open-eyed eye frames resemble a pair of reading glasses, with four LED's mounted in a circular pattern around each lense. This type of eye frame was developed by Ray Wolfe of Microfirm, Inc., and myself to incorporate light and sound entrainment into EEG/neurofeedback training. Particularly when training in the beta brainwave frequencies. These open-eyed eye frames allow the participant to receive light stimulation while their eyes are open, enabling the participant to observe the computer monitor during their neurofeedback training session. Research suggests open-eyed light stimulation increases the production of the desired brainwave frequency during neurofeedback training. Users have also found this type of eye frame to be beneficial while working or studying.

Lumasoft eye frames are the third type. From the outside, they resemble standard LED eye frames. But Lumasoft use small, white incandescent bulbs instead of LED's. A screen is mounted over the bulbs so users may keep their eyes open during sessions. If you have ever tried ganzfeld glasses and found the experience enjoyable, you will also enjoy the light stimulation generated by Lumasoft eye frames.

There are two drawbacks to the Lumasoft eye frames when compared to standard LED and open-eyed eye frames. First, the incandescent bulbs run hotter than LED's and second, Lumasoft eye frames are more expensive.

All three types of eye frames do basically the same thing: they blink. This on/off blinking is controlled by your light and sound instrument. How the lights in your eye frames blink is referred to as "phase".

Phase or how the blinking lights are combined can take several forms. The four most common are: standard in phase, standard out phase, independent light stimulation (ILS) in phase and independent light stimulation out phase.

Standard in phase light stimulation occurs when both the left and right lights are blinking at the same time. In phase means being in "synch". When this synchronization is being experienced, your ability to produce the desired brainwave is increased.

Conversely, standard out phase occurs when the left eye is on and the right eye is off, continuing in an alternating fashion. Just as out phase is the opposite of in phase, alternating light stimulation tends to have an inhibiting effect on achieving the desired brainwave.

Independent light stimulation (ILS) is like a visual binaural beat. For example, your left eye is receiving 12Hz stimulation and your right eye is receiving 18Hz. What this light stimulation is actually creating is an offset frequency of 6Hz, which is associated with the theta brainwave state.

The in phase and out phase of independent light stimulation means basically the same as the standard phases: being in synch or alternating. It is a little more complicated because different frequencies are being utilized. However, ILS phase techniques are very useful. For instance, by stimulating one eye with a high frequency and the other eye with a much lower frequency, you can actually activate or "excite" one hemisphere of your brain more so than the other. Or you may wish to activate both hemispheres simultaneously for accelerated whole brain learning.

What phase your light stimulation is set on greatly determines your session experience. Experiment with the different phases. Try short sessions combined with the various types of audio stimulation and make note of the mental, emotional and physical sensations you experience. Ask yourself which phase or combination of phases work best for you.

The duty cycle of light stimulation is for knowing how long the lights are on and off. So if the duty cycle is 60%, the lights will be on for 60% of the time and off 40%. Or if the duty cycle is 20%, that means the lights are on for 20% and off for 80% of the time. The higher the duty cycle, the brighter the lights appear. The lower the duty cycle, the sharper or more distinct the lights appear. Always vary the duty cycle when creating sessions as this will enhance the entrainment value and visual effects sensations of your session experience.

Intensity is how bright the light stimulation is. Light intensity is always user controlled. The beginning light and sound user should begin with the intensity at a low level. As more sessions are experienced, tolerance to light stimulation will increase. Gradually increase the light intensity as your tolerance increases. Remember: brighter is not always better, comfort is more important.

3. Tactile (Vibration) Stimulation.

Tactile stimulation uses physical vibrations to create a complete mind/body sensation. Much like biofeedback, light and sound is a "from the neck up" experience. Our emotional and mental states are of primary focus. But when tactile stimulation is incorporated into the light and sound process, not only are your emotional and mental states involved, but your physical state as well. This takes the "from the neck up" experience and elevates it to a "complete mind and body" experience.

Many of us are unaware of how our emotional and mental states can effect how we feel physically. Yet stress and anxiety can manifest themselves in physical forms. Illness is a prime example. As mentioned earlier, researchers and doctors believe 70-90% of illnesses are due to stress. Incorporating tactile stimulation with light and sound enables you to feel, as well as be mindful of, the sensations experienced during light and sound sessions.

I first experienced tactile stimulation in conjunction with light and sound in 1995. Utilizing a bed by Integrative Motion Systems with a Photosonix Nova Pro light and sound instrument, I began to explore what frequencies appeared to offer the most benefits, in terms of tactile stimulation. This research suggests that dual binaural beats in the 20Hz to

85Hz range were the most enjoyable and effective.

You do not need to invest an enormous amount of money to add tactile stimulation to your light and sound instrument. There are a variety of sound cushions available for around one hundred dollars and they are fully compatible with light and sound instrumentation. Do not be misled by vibrating mats. Make sure you purchase sound cushions that contain high quality transducers (bass "shakers" or speakers). Have two cushions, one for your back and one for your upper leg area. Applying the two cushion/two transducer stimulation will enable the dual binaural beats to gently guide your body into the same targeted state that your mind experiences.

Combining light, sound and tactile stimulation has become the ultimate experience in personal inner exploration. It is a very dynamic method for enhancing the general well-being for anyone fortunate enough to experience such a session.

IV. HOW TO USE YOUR LIGHT & SOUND INSTRUMENT

Reading the owners manual that comes with your light and sound instrument is always the recommended thing to do first. But if you are like me, it usually is second. Fiddling with the unit always has priority. So in an effort to have you do the recommended thing first, I will keep this short and informative.

A. How To "Quick Start" Your Light & Sound Instrument.

Let's begin with **sound only**. Since most sound only instruments have the same buttons and knobs, we will use the Photosonix 515 as our example. Do you wish to use small computer speakers or headphones? They both connect to the 515 the same way. Take the male plug of either the speakers or headphones and connect it to the 515 jack located on the right side of the unit marked with a circled "S". Do not connect it to the 515 jack marked "EXT". Next, take the stereo patch cord that comes with your system (one long cord with male plugs at each end) and connect one end to the headphone jack of your CD or walkman and the other end to the 515 jack located on the right side marked "EXT" and insert whatever cassette or CD you intend to listen to. Take the AC adapter and connect the female end to the jack located on the left side of the 515 and insert the AC plug into a standard electrical wall outlet. You are now set up and ready to go.

Your biggest decision is deciding which session you want to listen to first. Session A1 (Energize) lights up when you press the "ON" button, so let's listen to B3 (continuous/learning). To get there, just press "SELECT" until the red LEDs of both the "B" and "3" are glowing. Now press the "PLAY" button. Adjust the volume by turning the thumb wheel located on the side of the 515 that is below the "SELECT" and "PLAY/STOP" buttons (the other "S" that is circled). Next, adjust the volume of your cassette or CD by adjusting your walkman or CD player's volume control. The audio tones generated by the 515 do not have to be played loud to be effective. Keep the audio tones at a volume that is comfortable for you. Have the accompanying music or spoken word at a higher volume. Guess what? You are now experiencing a 515 session. Enjoy!

Now for quick starting your **light and sound** instrument. The following instructions will show you how to immediately enjoy the Photosonix Nova Pro. Should you own a different light and sound instrument, no need to worry, all light and sound instruments work pretty much the same way and these instructions will get you going with almost every light and sound unit that is available.

First you need to press the power (PWR) switch. This obviously turns on your system. You will see P01 displayed, which is the program that is ready to run. Now press the SELECT button. Each time you press this button, notice how a different light will be on. This tells you which parameter is being displayed. When SELECT reaches the furthest parameter on the right, there are three selections to choose from. Pressing SELECT once more will return the display back to PROGRAM. By pressing either the UP or DOWN ARROWS, you can change the values of the parameter that is displayed. Try this: Press the POWER switch. Notice the red LED under PROGRAM is on. Refer to the list of programs in your owners manual and decide which program you would like to experience. By pressing the UP ARROW, go to that program. Now press SELECT and go to VOLUME. By pressing either the UP or DOWN ARROWS, find the volume that is comfortable to you. Do the same thing with INTENSITY. This parameter controls the brightness of the lights. Be nice to yourself-adjust the INTENSITY with your eyes closed and while wearing the eye frames. And start with the brightness at a low level. Build up your tolerance to the flashing lights gradually. Keep the INTENSITY setting at a comfortable level!

Press SELECT until the red LED under PROGRAM is on. Now press ENTER. There will be a ten second countdown displayed before the program begins. This will allow you enough time to get settled in comfortably, with your headphones and eye frames on just right. Enjoy your session!

Almost every owners manual that I have seen has simple directions for quick starting their light and sound or sound only units. It is a good idea to refer to your manual should you be unclear on how to properly use your instrument.

B. Helpful Hints.

-Always read your light and sound/ sound only owners manual. It is the best way to maximize your session experiences.

-Always have the volume at a comfortable level. Blasting the audio tones will not make a session more effective.

-When wearing eye frames, keep your eyes closed and have the brightness set at a level you find enjoyable. The more you use your instrument, the more tolerable to brighter intensities your eyes should become. Like audio tones, you do not need to have the brightness set at the maximum level for it to be effective for you.

-To achieve the best possible results, use your system on a daily basis for the first 30 days. Refer to the next chapter (30 Day Light & Sound Program).

-Wear clothing that is comfortable and loose fitting while doing sessions.

-Make using your system a routine. Try to set aside a specific time and place for you to use your system.

-Have a goal in mind before beginning a session and say it verbally to yourself. Maybe it is a problem you need to solve, an idea you need to organize or some stress that needs to be released.

-Try to draw a mental picture of the goal you have in mind. Try to make it as detailed as possible. Can you picture the results? During the session, try not to drift from this picture. Try to remain focused on your goal and mental picture. Should you find your thoughts drifting, just refocus.

-Keep a blanket or comforter handy. Blood flow to your extremities is reduced during deep relaxation and you might find yourself a bit chilled during the session.

-Be aware of how you are breathing. Try to have your breaths come in a deep, regular fashion. Try this: Inhale (count 1,2,3,4), Hold (2,3,4), Release (1,2,3,4) or mentally say to yourself as you breathe: Quiet/Cope/Focus.

-Keep a log or journal of your session experiences. After a session, remain still and listen to any thoughts that come into your mind. Answers do not always immediately surface after a session. Sometimes they surface hours or even a few days later. Be conscious of this while you go about your normal daily routine.

-To determine what works best for you, become aware of your energy levels, sleep patterns and stress levels over the course of the first four weeks.

-Do not be overwhelmed by your light and sound instrument. The more you use it, the easier it becomes to use. Should you ever have a problem or question, never hesitate to call the dealer you purchased it from or the manufacturer directly.

-Please remember: When I first started using a light and sound instrument (an IQ-III in 1986), there was no one I could call unless I wanted to make a long distance phone call. I still have a difficult time programming my VCR. So if I can figure this technology out, believe me, you can too!

C. Adding Music Or Spoken Word To Your Light & Sound Sessions.

Words can never fully describe the sensations you will experience when you combine your favorite musical selection with a light and sound program. Try it and you will understand. All you need is a stereo patch cord, which should always be included with your purchase of a light and sound instrument.

What you do is connect one end of the stereo patch cord to the external jack of your light and sound unit and connect the other end of the patch cord to the headphone jack of your cassette walkman, boombox or portable CD player. You can adjust both the audio tone level and music through your light and sound instrument. You will find it easier to adjust the volume of the music through your tape/CD player after that.

Do you enjoy rock and roll? Try listening to an energize (high beta and gamma frequencies) session to the accompaniment of the Stones or AC/DC. Audiogasm by Thomas Michael can also be an excellent choice. Almost all fast-paced music combined with high frequency sessions can get you proverbially "wired".

Rock not your style? Heart beating too fast already? Need to slow down and shed that stress? Go the other way. Try listening to slower-paced music with sessions in the alpha and/or theta frequency range. After experiencing this type of session, you won't need to swallow that valium. You will feel naturally relaxed. Try this method with your favorite selection of "soft" music or environmental sound effects tape. Like I said, the experience is very hard to describe.

What if you are dealing with some issues and you have a self-help tape to assist you in overcoming your problems? Combine that tape with a light and sound session in the alpha and/or theta frequency range. By using light and sound while listening to that tape, you should find it easier to accept the information that tape is suggesting as compared to just listening to the tape by itself. Why? Because utilizing light and sound will take you to the level of consciousness where your mind is more acceptable to new suggestions and change.

Are you a student or are you facing a promotional exam at work? What do you do with all those notes you've been taking? Try this: In your own voice, make an audio cassette based on your notes. Listen to this cassette in conjunction with an alpha or theta session, or a combination of both. Whichever works best for you. I have had too many clients tell me alpha works best or theta works best, for me to be any more specific. The important thing is you will know what works best for you. And you should know within a matter of one or two days. You will realize you have increased memory retention of what you need to know. Plus you should experience less anxiety associated with your upcoming exam. But if you procrastinate or put off studying until the last minute, don't even bother trying this method. Instead, skip the exam and by a self-help tape (see above paragraph).

Sound only instruments can also be combined with the methods described above. They usually connect the same way as a light and sound instrument does. Some models even go one step further. Not only do they come with a stereo patch cord, they also come with an auxiliary cord that has one male connector on one end and two male connectors on the other. This way the single end connects to the sound only unit and the double or "Y" end connects to the left and right audio channels of your home or office audio stereo system. This negates the need to sit still while you are doing a session and allows you to move around, for whatever reason, and still receive the benefit of audio frequency stimulation.

D. Understanding What Built-in And Downloadable Sessions Are.

Selecting a session is simple. Your owners manual contains a complete list of all built-in sessions, plus instructions on how to either custom design or download others.

You should first ask yourself which type of session you feel like experiencing. The session list will contain such types as energy, relaxation, creativity, meditation, learning and sleep. Decide which type you feel in the mood for and check which sessions are available within each type.

This list will also tell you what frequencies are targeted and the intended general effect. Since we are all different, that is, no two individuals think exactly alike, the general effect of each session will vary on an individual basis. In general, however, if a session deals with relaxation, all individuals should expect to experience varying degrees of relaxation and calmness. The same idea also pertains to the other types of sessions.

Session titles are usually chosen by the session designer and are based on the programmer's experience with the session and also the experiences of test subjects. Just keep in mind that since we are all different, we will all be affected differently by each session. Therefore, the best way to determine which sessions work best for you is to experiment with each session title within each session category.

Open your owners manual to the section that lists the sessions. In the first column the sessions are broken down into types, like energize, relaxation, meditation, entertainment, etc.. Then you have the session numbers followed by their length. Next comes the titles of each session, their frequency range, type of tone used and a brief definition of what each session tries for.

Here is the easiest way to figure out which one you want to use. Find the type of session you feel like doing, then go straight to the frequency range. Of course you already know the different brain states and general characteristics of each state because we have covered that earlier. So for example, say you need to relax. You go to the relaxation sessions. How deep do you want to go? The frequency range will tell you how deep the session goes, not the title of the session or the brief definition that follows. Those are informative, but it is easier to just look at what frequencies the session targets.

Two sessions target the same frequency? Check which type of tone is used. Length of session is also important. Decide what length you want. Remember, always listen to the complete session. Do not listen half way through then go out and drive the freeways. That is not conducive to a healthy lifestyle. Besides, most light and sound units give you the opportunity to either lengthen or shorten the time of each session. Also, most allow you to change the type of tone for each session. Over time, you will probably try all of the sessions, you will know the sessions you like the most and you will stop referring to the owners manual when you go to use your system. That's okay. That says you know what your system has to offer and how you can manipulate the parameters to suit your specific needs.

Many sessions target more than one brainwave category. Say the purpose of doing your session today is for creativity. You are working on a project and you need some fresh ideas, but you are facing a deadline for getting it completed. Do you do a short creativity/theta session for some quick inspiration or a half-hour creativity/energize session to conjure up fresh ideas and simultaneously produce the energy to write it down really fast? Try the short creativity/theta session. Why? Because at sessions end, you will be back at your normal low beta state, feeling relaxed and fresh and hopefully full of new ideas. And your focus will also be enhanced. Being in a relaxed and focused frame of mind should enable you to write what needs writing in a fashion that meets your deadline. Your mind should also possess the clarity to carry you through to project's end.

On the other hand, how late were you up last night? Do you need the alertness associated with high beta/energize stimulation? Can you spare the extra time doing a longer session and still meet your deadline? Say you need the alertness associated with the energize session but lack the half-hour to do the session. Why not take the best from both sessions? All you have to do is shorten the length of the creativity/energize session. If it is 30 minutes, make it 15 instead. Just make sure at session's end that you sit down to finish your project and not run out to hit some tennis balls.

That is just an example of how you can adjust the built-in sessions of your instrument to suit your specific needs. When doing a session, see if this statement holds true for you: When experiencing a beta or gamma session, the colors are predominantly black, white and grey with a lot of intricate patterns. When doing alpha, theta and delta sessions, there are more colors, lots of green, blue, red, yellow, etc., and there are more swirls and blends rather than patterns. I am not sure why that is. I do know all those visuals are produced by your mind. And after performing thousands of light and sound sessions on clients, friends and associates, I have found this to be the norm rather than the exception. But as for why that happens, either refer to the Suggested Reading section or ask your local neurologist. They usually know everything.

E. Designing Your Own Sessions.

As of yet, not one institution of higher learning offers a doctorate in light and sound technology. So who are the designers of light and sound sessions? People just like you. All they possess that you do not is experience. But with patience and a few ounces of creativity, you too can learn the recipe for creating effective and beneficial light and sound sessions.

This is what you need to get started: you need to understand the interaction between brainwave frequencies and the states of mind that they are associated with; and you need to take the time to listen to the sounds created by individual frequencies. What sounds good? What sounds completely annoying? Combine and refine.

Once your session is finished and you are happy with the results, ask others to try the session to see if their experiences are similar. When approaching them, say as little as possible. Not to appear flighty, but to give them as little of preconceived notion as possible. After all, you would like their honest reactions. Remember, you are not looking for their analysis of the session, but rather how differently they feel after the session as compared to how they felt before.

Now, how do you actually structure a light and sound session? There are six primary ways to manipulate frequencies when constructing a session. They are: using a ramp, jump, loop, microloop, holding a frequency and stairway motion.

When you change a frequency or pitch in a smooth and gradual fashion, either up or down, you are using a ramp. For instance, over a two minute period, you gradually increase 4Hz to 8Hz. When you use a jump, you are increasing or decreasing from a given frequency at random.

Example of "Ramping" in session design:

7Hz 8Hz 9Hz 10Hz

Utilizing gentle frequency increases (or decreases).

Example of "Jumping" in session design:

12Hz 8Hz 24Hz 14Hz

Movement of frequencies are random.

Repeating a group of selected frequencies or pitches in a given sequence is referred to as a loop. Microloops are shorter in time duration, usually each step being less than one second.

Example of "Loop" design when creating sessions:

(7Hz, 12Hz, 17Hz) (7Hz, 12Hz, 17Hz)

Frequency pattern repeats at least twice. Note: Same procedure for microloops except each step is less than one second.

When you are holding a frequency, you are remaining at a specific frequency for a specific period of time. By contrast, utilizing a stairway motion means that you are keeping the numerical distance of the steps constant throughout the cycle.

Example of "Holding a frequency" in session design:

10Hz 10Hz 10Hz 10Hz

Utilization of same frequency (keeping it steady).

Example of "Stairway" design when creating sessions:

8Hz 10Hz 12Hz 14Hz

Movement of frequencies at same intervals.

These methods of constructing sessions should be more fully explained in your owners manual. You can also refer to the Glossary and Suggested Reading sections for further information.

Let personal preference guide you as far as session length. Many believe thirty minute sessions are most advantageous. Others believe seven to twelve minute sessions work best. Since you are the one designing the session, you should be the one best able to make the decision. If it sounds good, if it makes you feel good, then go with it!

Every session should have three sections: the beginning or induction period, the middle or core period and the ending, where returning to your normal waking state is highly recommended.

An induction period can be long in duration. Either holding a frequency or gradually ramping up or down to guide the listener towards the core. This longer induction is sort of like a "settling in" period or "calm before the storm", depending upon the type of session you have in mind.You may also choose a short induction or none at all. Save time and just zoom right to the core. It all depends on what you prefer.

The core or middle section is really the "meat" of your session. First, know what mind state you are aiming for and target frequencies that are associated with that state. Second, decide how long you want this section to run. To determine which frequencies to target for each mind state, refer to the general brainwave categories starting on page 10.

When designing a gamma, beta, alpha, theta or combination, you should have a predetermined core length so your ending section can guide the listener back to their normal waking state. But for a delta or sleep session, where there is no need for returning to a low beta state, your core section can gradually fade out with no need for an ending section.

What matters most when constructing the end section is how you want the person to be left feeling. Ending a session in high beta before physically working out can be ideal. But if the listener is experiencing a relaxation session before driving the freeway, it would not be nice to end the session in theta or alpha. Ending the session in low beta would be much more safer.

For whatever reason you have for designing a session, be it for energy, creativity, relaxation or just plain fun, view the design as if you were writing a song or a story. How do you want them to feel as they listen to the session? How do you want them to feel afterwards?

As you become familiar with the different ways of constructing sessions, you will acquire little secrets that can be used in designing most any session. All you need is patience, the opportunity to experiment and confidence in your ability to create. And there are plenty of avenues for learning more. Just consult your light and sound owners manual and the Suggested Reading section.

V. BEGINNING YOUR JOURNEY

A. The 30 Day Learning/ Relaxation Light & Sound Program.

When I began my education into light and sound technology, the consensus amongst this field's leading researchers was that one 30 minute (minimum) session per day was the most beneficial way for an individual to use their light and sound instrument. That was over ten years ago. Recently, I attended a conference where this field's leading researchers suggested two, perhaps three 10-15 minute sessions per day is the most beneficial way. As much as I respect the medical professionals and research leaders that have done so much in getting this technology accepted by the medical community in general, I do believe they are as likely to change their opinions in the future as they have done in the past. Therefore, I will say to you exactly what I say to my clients. Everyone is different. Everyone's daily routine is different. You are the one who knows what works best for you. That is why, when you begin to use your light and sound instrument, you do it daily and in a way that is convenient for you. Let this 30 day program be your guide and utilize the users log to write down your personal session experiences.

Ask yourself if it is easier for you to do one long session per day, or is it easier to do two or three shorter sessions over the course of the day? Which approach better fits into your schedule? You shouldn't have to worry about making time to do sessions. That just isn't very relaxing!

Whether you do one long or two to three short sessions per day, remember that a key to your success is consistency. By keeping to your schedule, in a short time you should begin to feel a reduction in stress, an improvement in your ability to relax and a general overall feeling of good health. You will begin to develop control over the daily stressors that influence your mental, emotional and physical being. Keep positive thoughts in your mind.

Day 1: Spend 30-40 minutes listening to an alpha relaxation session on your light and sound instrument. You can combine soft-paced music or environmental sounds to enhance the effect. At session's end, take 10 minutes to allow your mind to file away the positive thoughts you experienced. Write these thoughts down in your daily log.

Day 2: Follow the same procedure as with Day 1. Remember to write your thoughts in your daily log. Visualize yourself being completely free and weightless as you learn to cope with daily stressors and challenges.

Day 3: Spend another 30-40 minutes doing light and sound. Still utilize alpha frequencies, but select a different session(s). After doing your session in alpha, choose an energize session that is 15 minutes in length. You should begin to notice subtle changes, like feeling more alert, but in a relaxed kind of way. If you find going to sleep at night is becoming a problem, use a sleep session on your light and sound unit to help you fall asleep. Do you feel more control over anxiety and any pain you may have been experiencing? Are you able to tolerate situations in a more positive fashion? Remember to use your daily log.

Day 4: Again spend another 30-40 minutes doing alpha session(s) on your light and sound unit, followed by a short (10-15 minutes) energize session. By now you should be feeling more in control of your life. If you don't, no need to worry, that feeling will be happening soon. Try explaining your feelings of calmness in your daily log.

Day 5: Listen to the same alpha/relaxation program as you have been for the same amount of time. If needed, remember to utilize a sleep program before bedtime. By now you should be aware of less muscle tension in your neck and shoulders. Thoughts and ideas are becoming more organized for you. Getting uptight or anxious is no longer a concern for you. You are in control and you feel good about it.

Day 6: Doing your 30-40 minutes of alpha light and sound session(s) is now part of your normal daily routine. Your breathing has become slower, you feel more relaxed and your muscles are no longer as tight as they have been for quite some time. You deal with stress more quickly now because you can feel it when your neck and body muscles become tight. You recognize the signs of stress and you deal with it. You have found a way to cope with daily stressors and you no longer fear them. Remember to write about your experiences and sensations in your daily log.

Day 7: Same procedure. Visualize feeling happy and positive with life. You have a feeling of confidence with greater self-control. You are more relaxed and calm, yet mentally alert. Your ability to focus seems more pronounced. Others begin to notice a positive change occurring just by the way you look.

Week 2: Explore deeper frequencies. Choose 30-40 minute (total) session(s) that take you into alpha and theta. Remember to use your daily log to record new sensations and experiences. Stress is no longer a part of your daily life. You are in control of your life.

Week 3: Continue your 30-40 minute daily session(s). Utilize sessions in both alpha and theta. Find session(s) in these frequency ranges that give you a positive experience and record your feelings in your daily log. You are motivated. You know you can listen to a sleep session at bedtime if needed. You have a sense of general well-being.

Week 4: Spend 30-40 minutes doing your daily session(s). You realize the sensations of being in alpha and theta. It is as if you can now achieve these states at will. You want to explore different brainwave states and the sensations that await you at each new level. Go for it. Record your experiences in your daily log. Notice the improvement in your attitude. Now is the time for quality mind exploration.

B. Daily Journal For Describing Personal Session Experiences And Sensations.

Changes do occur within you as light and sound becomes part of your daily routine. But these changes are very subtle. By and large, the realization that certain things are different happen after the fact, not during, Using this daily log for recording your daily experiences and sensations will help you realize that these changes are occurring. In maintaining a daily record, you benefit by having the ability to review what you have previously written and are able to compare those feelings with what is happening to you now. So use this section. In the days ahead, you will be glad you did!

What do you write down? What types of things are important enough to keep track of? Here is a list of questions that can help you decide. You do not need to use them all. Perhaps there are others that are more pertinent to your situation. You make the decision. After all, it is your daily log.

-What session(s) did you use? For how long? What kind of patterns and colors did you see?

-On a scale of 1-10 (1=lousy/10=excellent), how would you rate the following levels both before and after doing a session: Your stress level?

Your energy level? Your ability to focus?

-In what ways do you feel differently after your session compared to before your session physically, emotionally, mentally? How long do these feelings/sensations last?

-Were you able to remain focused on your goals during the session or did your mind drift? If your mind did drift, were the thoughts pleasant? Were you able to refocus?

-Did you have any new ideas? Any "profound" revelations?

-Is your memory improving? In what ways? Do you sense improvement in your ability to concentrate?

-Do you feel more centered? Are you more organized?

-If you suffer from chronic pain, is the pain decreasing? Do you suffer from muscle tension, tightness or headaches? Are they decreasing in severity?

-Do you experience mood swings? Do you notice them decreasing in frequency and/or intensity?

-How are you sleeping? Are your dreams lucid?

Your Personal Daily Journal:

Day 1:

Day 2:

Day 3:

Day 4:

Day 5:

Day 6:

Day 7:

Day 8:

Day 9:

Day 10:

Day 11:

Day 12:

Day 13:

Day 14:

Day 15:

Day 16:

Day 17:

Day 18:

Day 19:

Day 20:

Day 21:

Day 22:

Day 23:

Day 24:

Day 25:

Day 26:

Day 27:

Day 28:

Day 29:

Day 30:

Additional comments:

VI. CLOSING THOUGHTS

You now know how to use your light and sound instrument. You realize you have given yourself a very valuable gift. One that will last you a lifetime. You are aware of your different states of consciousness.. And you are aware of how powerful accessing these states can be on your mental, emotional and physical being.

This lifetime gift is knowing how to relax Learning how to relax is the basis of light and sound technology. You now know how, quickly and effortlessly. Use your power to relax as the starting point for reaching your true potential.

As you progress, as you access new levels of your consciousness, remember how you felt when you began your journey. Keep track of how you feel, write down your new sensations and experiences. Because you can reach your true potential. And when you look back, when you read what you have written, you will realize what a wonderful path you have chosen. And you will be truly amazed.

Now I would like to take a moment to address an issue that has received very little attention from the leaders in the light and sound community. I have always felt that the visually impaired have been neglected when the mention of light and sound is discussed. But how would you react if suddenly your vision began to deteriorate? Would you feel anger? Would you be afraid? Do you think learning the Braille keyboard is easy? There is a lot of frustration involved. And an incredible amount of stress. Okay, granted the pulsating light aspect of this technology would have little effect, but dual binaural beat sound generation sure helps. I have participated in the Braille Institute's Open House and have heard their concerns. I have spoken about these concerns with Chris Oliver on radio (KPFK Challenge talk show).

To illustrate how sound generation can assist the visually impaired, I would like to share the following experience that happened to one of my visually impaired clients. Due to the deterioration of her eyesight, Laura Riggs had to return to school because she could no longer fulfill the responsibilities associated with her profession. Returning to school after twenty-five years was a real undertaking for her. Laura experienced difficulty with retaining new information which in turn caused more stress. Her grade point average was at 2.0 or a "C". She then began using the 515 sound only instrument. Her anxiety level declined and her test scores improved. She began to retain the information being read to her on tape. And her grade point average increased to a 3.2 or "B+". And at this moment, Laura is completing her final semester at a major California University.

Hopefully, research can increase concerning sound generation and the visually impaired. And perhaps those with hearing deficits can receive attention in regards to research exclusively utilizing pulsed light stimulation. After all, we all experience various forms of stress.

In sponsoring learning/relaxation programs, I have had the opportunity of using several light and sound models designed by various manufacturers, including InnerQuest, Microfirm, Synetic Systems and MindGear. They have all proved to be effective mind entrainment instruments.

Since 1995, I have utilized Photosonix (Microfirm) equipment exclusively. There have been a number of reasons for this. As a light and sound dealer, I appreciated the willingness of Ray Wolfe (CEO of Microfirm) to listen to what was needed to successfully carry out my programs. He assisted in designing specific sessions to meet the needs of the participants. He created instruments that are easy to understand and use for the beginning light and sound user. Ray even worked with Chris Oliver and myself to design a sound only instrument for people who are visually impaired or unable to experience light and sound due to possible adverse effects.

Because a sound only unit could not be as effective as light and sound in terms of entrainment capabilities, if existing sound stimulation was utilized, the three of us worked together to create a new form of audio stimulation that became known as dual independent binaural beats. This form of audio stimulation has proven so effective that all Photosonix instruments now possess dual independent binaural beat capabilities.

By no means should my decision to carry Photosonix instrumentation exclusively be construed as a slight on the other light and sound manufacturers. The original reason for my doing so was geographical. Our businesses are both in the greater Los Angeles area. Working with Ray is obviously much easier than trying to work with someone in Washington or Ohio.

Robert Austin (CEO of Synetic Systems), goes to great lengths to meet the needs of his light and sound dealers. I know from personal experience. And George Szeless (CEO of MindGear) was the manufacturer who offered me assistance when I began establishing light and sound programs in the law enforcement community. As I have said earlier, every light and sound manufacturer that I have met has always been willing to listen and to help when they can. Remember this should you ever have a question and cannot find the answer.

Until now, I have always kept the learning and relaxation programs described in this book confidential. After all, people have a right to keep what is happening in their personal lives private. Plus, when working with public schools and government agencies, for example, there are many underlying issues (mainly politics) that could surface should the general population take notice. Why? Because doctors and psychologists are expected to prescribe pills, not cost-effective instruments, to help people in need. And teachers are expected to follow a lesson plan, regardless of their class size and budgetary constraints. But mainstream America is wising up. There are a great many tools and techniques out there that are proving to be safer and more cost efficient than what has been termed "acceptable" by professional medical practitioners.

Some of my programs will remain confidential. But I wrote this book because so many of my clients thought I should. They believe in this technology. They purchase a unit for a specific purpose and find new ways of incorporating it into their lifestyle. Be it for business, pleasure, academic enhancement or for overcoming pain, they are the ones who share their experiences and they are the reason why this technology is gaining acceptance. To all my clients, you have taught me a lot about light and sound. Thank you!

****

In closing, I would like to ask all of you who own pets for your assistance. I believe sound frequencies could be of benefit to these members of our family, but have experienced difficulty in having this idea accepted by veterinarians.

This idea came to me when I had to leave one of my kittens overnight at the vets for observation. Picture this: One large room of people's pets that are in cages. All of them are stressed from being away from their families and being placed together in this room. Many do not feel well. Others are recovering from operations. They all feel alone. It is nighttime and all the lights are out. Hopefully an attendant occasionally checks in on them.

There are sound sessions that could help them relax, perhaps even fall asleep.

Now if relaxing is conducive to our well-being, aiding us in our response to medical treatment and recovery, is it too hard to believe that our pets would experience the same benefits? Please give this idea consideration, for your own pets

VII. REFERENCES CITED & SUGGESTED READING LIST

A. Selected References:

Atwater, F.H. (1988). The Monroe Institute's Hemisync Process: A Theoretical Perspective. Faber, VA.: Monroe Institute.

Budzynski, T.H. (1991). Hemispheric Assymmetry and Rest. In Suedfeld, P., Turner, J.W. Jr., & Fine, T.H. (Eds.), Restricted Environmental Stimulation. New York: Springer-Verlag.

Ditson-Sommer, R. (1995). Neurolearning Technology. Phoenix, AZ.:OPNET, Inc.

Foster, D.S. (1990). EEG and Subjective Correlates of Alpha Frequency Binaural Beats Stimulation Combined with Alpha Biofeedback. Ann Arbor, MI.: UMI. Order No. 9025506.

Hoiberg, B.C. (1989). Introducing Your Brain To It's Higher Possibilities. Excerpt from presentation given at 1989 PSI Conference in Charlottesville, VA.

Hutchison, M. (1990). Special issue on sound/light. Megabrain Report: Vol.1, No. 2.

____________ (1992). A Short History of Light/Sound Technology. Seattle, WA. : Synetic Systems, Inc.

Niedermeyer, E. and Lopes Da Silva, F. (1982). Electroencephalography: Basic Principles, Clinical Applications and Related Fields. Urban & Schwarzenberg.

Oster, G. (1973). Auditory Beats in the Brain. Scientific American, #229, 94-102.

B. Suggested Reading:

If you are on the Internet, try http://www.altavista.digital.com and type in light and sound. There are many other search engines, such as Yahoo, but Altavista will supply you with hundreds of sites filled with information concerning light and sound technology.

Or you may wish to refer to the following books and authors:

Accelerated Learning:

Anything written by Rayma Ditson-Sommer, Ph.D., of Phoenix, AZ. Dr. Sommer is the creator of BHINS, and has spent years helping individuals with learning disabilities.

"Superlearning" by Sheila Ostrander and Lynn Schroeder (Delacorte, 1979).

"Super-Memory: The Revolution" which is by the same authors (Caroll & Graf, 1991).

Anchoring:

"Trance-formations" by Grinder and Bandler (Real People Press, 1981).

"Unlimited Power" by Anthony Robbins (Fawcett, 1986).

Focus:

"The Open Focus Handbook" by George Fritz, Ed.D., and Les Fehmi, Ph.D. (Publisher Unkown).

"Focusing" by Eugene T. Gendlin, Ph.D. (Everest, 1978).

Meditation:

"Seeking the Heart of Wisdom: The Path of Insight Meditation" by Joseph

Goldstein and Jack Kornfeld (Shambala, 1987).

"Full Catastrophe Living" by Jon Kabat-Zinn, Ph.D. (Delacorte, 1990).

"A Gradual Awakening" by Stephen Levine (Anchor/Doubleday, 1979).

"Zen Mind, Beginner's Mind" by Shunryu Suzuki (Weatherhill, 1986).

Neurotechnology:

"Mega Brain Power" by Michael Hutchison (Hyperion, 1994).

"Awakening Mind I: Creating Sound and Light Sessions on Advanced Programmable Mind Machines" by James Mann (Enlightened Enterprises, 1992).

"Mind Machine Buyer's Guide" by Neuro-Technology Research (Neuro-Technology Research, 1994).

Peak Performance:

"Health Psychology: Integrating Mind & Body" by George D. Bishop (Allyn & Bacon, 1993).

"Better Health With Self-Hypnosis" by Frank S. Caprio, M.D. (Prentice Hall 1985).

"Ageless Body, Timeless Mind" by Deepak Chopra (Random House 1994).

"Creating Health: How to Wake Up the Body's Intelligence" by Deepak Chopra (Houghton Mifflin 1995).

"Being There: Putting Brain, Body and World Together Again" by Andy Clark (Mit Pr/Trd 1996)

"Imagineering for Health" by Serget King and Serge Kahili King (Theosophical 1981).

"In the Zone: Making Winning Moments Your Way of Life" by Ray Mulry, Ph.D. (Great Ocean 1995).

Relaxation:

"The Mind/Body Effect" by Herbert Benson, M.D. (Simon & Schuster, 1979).

"Minding the Body, Mending the Mind" by Joan Borysenko (Bantam, 1988).

"The Book of Floating" by Michael Hutchison (Morrow/Quill, 1984).

"The Fine Arts of Relaxation, Concentration and Meditation" by Joel Levey (Wisdom Publications, 1987).

"The Complete Guide to Your Emotions and Your Health" by a compilation of researchers, edited by Emrika Padus (Rodale Press, 1992).

"You Can't Afford the Luxury of a Negative Thought" by John-Roger and Peter McWilliams (Prelude Press, 1988).

Rescripting:

Any articles/books by Thomas Budzynski, Ph.D.

"Programming and Metaprogramming in the Human Biocomputer" by John C. Lilly (Julian, 1972).

"Software for the Mind: How to Program Your Mind for Optimum Health and Performance" by Emmett Miller (Celestial Arts, 1987).

"Do It!" by John-Roger and Peter McWilliams (Prelude Press, 1991).

Self-Hypnosis:

"Bodymind" by Ken Dychtwald (Pantheon,1977).

"The Answer Within: A Clinical Framework of Ericksonian Hypnotherapy" by Lankton and Lankton (Bruner/Mazel, 1983).

"Self-Hypnotism" by Leslie LeCron (Prentice-Hall, 1964).

"The Psychobiology of Mind/Body Healing" by Ernest Rossi (Norton, 1986).

Visualization:

"Creative Visualization" by Shakti Gawain (Publisher Unknown).

"Seeing with the Mind's Eye" by Samuels and Samuels (Random House, 1975).

VIII. GLOSSARY

Activation frequencies are used in session design to "turn on" or guide the participant to specific states of awareness, i.e. beta, theta, etc..

Alpha brainwave frequencies are associated with being relaxed, yet mentally alert. The frequency pattern ranges between 8-12Hz.

Beats are amplitude variations that result when pure tones spaced closed together in pitch are mixed.

Beta brainwave frequencies are associated with are normal waking state. Beta frequencies range between 13-34Hz.

Binaural beats result when pure tones at different beats are simultaneously heard by both ears. For instance, say one tone of 200Hz is heard predominantly by the left ear. Another tone of 190Hz is heard predominantly through the right ear. These two pure tones are heard and processed separately, depending upon their arrival time, by our auditory center. If these two tones are of the same pitch, the resulting sound will be perceived as being inside our head. But if both tones have slightly different pitches, the resulting effect will sound more like a back and forth movement.

Biofeedback is a process by which participants must generate the targeted brainwave with no outside stimulation.

Biofeed-in is another term used to describe how the entrainment process of light and sound stimulation is performed. The opposite of biofeedback.

Brainwaves are the electrical wave patterns generated by our brain. Our brain contains roughly 100 billion neurons that function by generating and passing these electrical signals. The sum of all these electrical signals and their activity can be recorded and observed via an EEG instrument. These brain waves appear as irregular yet repetitive "wavy lines" that are a mixture of frequencies ranging from below 1Hz to more than 40Hz. The predominance of a certain brain wave frequency is what determines what brain state or state of awareness an individual is experiencing at any given moment.

Chakra frequencies are associated with our inner power, referred to in eastern religions as the kundalini. These frequencies can activate certain centers along the base of the spine upwards to the brain. These frequencies are not the natural sounds associated with the activation of chakras, but do an excellent imitation, resulting in very beautiful and powerful experiences.

Chords are tone combinations that we want to sound harmonious. This happens when the fundamental pitches of the tones occur in ratios of small integers. For example, a 2/1 ratio is an octave, 5/3 is a sixth, 3/2 is a fifth, 4/3 is a major third and a ratio of 6/5 is a minor third. Most of us hear tones that are combined closer than a minor third as dissonant. When the frequencies of tones are combined even closer, beats occur.

Consciousness management technique refers to the overall process of light and sound technology.

Consciousness self-regulation technique refers to the overall process of biofeedback.

Delta brainwave frequencies are associated with sleep. These frequencies range between .5-3Hz.

Downloading is the process of transferring sessions from a PC computer to a light and sound/sound only instrument.

Dual binaural beats occur by running two binaural beat generators simultaneously. Two tones plus two binaural beat offsets are heard by both ears, for a total of four tones and four offsets (the two left offset tones against both right offset tones). The results produce very complex and dynamic frequency variations that may seem confusing, but sounds really great. Independent dual binaural beats can only be found on Photosonix light and sound/sound only instruments.

Duty cycle impacts the sensation of light and sound stimulation. It refers to how long the light or sound stimulation cycle is on. For instance, if the duty cycle is set at 30%, that means the stimulation is on for 30% of the cycle and off for 70%.

Entrainment is the method light and sound/sound only instruments use to gently guide your mind to desired states of consciousness.

Euphoria sessions target the activation of the left brain hemisphere while leaving the right brain hemisphere inactive. The approach in session design is aimed at producing a sense of well-being.

Frequency refers to the primary rate of pulsed light (visual) and pulsed tone (auditory) stimulation produced by a light and sound session. It is how fast the lights and sound are at one moment on, then at the next moment off. In other words, how fast each form of stimulation is pulsating. In terms of binaural beats, frequency refers to how often the amplitude is being modulated. Light and sound sessions utilize the same frequency range as the brain, i.e. brain waves.

Frequency Following Response refers to the brain's natural tendency to mimic or resonate the same frequencies generated by outside stimuli. This is the process light and sound instruments utilize for entraining the brain.

Gamma brainwave frequencies are associated with being highly motivated and mentally alert. These frequency patterns start at 35Hz and increase.

Ganzfeld in German means "all-field". Refers to gazing into a snowy-white background as a technique for achieving altered states.

GSR stands for galvanic skin resistance. It measures the variations of your skin's sweat gland and pore size activity, both of which are controlled by the sympathetic nervous system. It is a measurement of your physical reactions to states of stress and relaxation.

Harmonic frequencies are basically multiples of each other. It is believed these frequencies act more in agreement or unison with each other than other frequency combinations.

Harmonic multiple frequencies are the result of multiplying a number by a lower offset frequency to produce a higher offset frequency. It is a technique used in designing binaural beat sessions.

Hertz refers to cycles per second, or how often something repeats itself within each second. Named for Heinrich Hertz (1857-94). Hertz is the primary measurement of sound, but is fast losing popularity to amplitude.

Integration in regards to light and sound session design refers to utilizing frequencies in specific brain states that aid the participant in accepting and incorporating what surfaces into the conscious mind during session experiences.

Intensity is how bright the light stimulation is.

Light frames are eye frames, i.e. eye wear or eye goggles with light sources mounted in the front inside portion of the frame. These are the pulsating lights that make up the light stimulation of a light and sound session. The lights themselves can be LED's, which are color-emitting diodes or white incandescent light bulbs. The LED light frames are utilized with eyes closed because the eyelids actually diffuse the lights and heighten the session experience. White incandescent light frames have a translucent mask between the light bulbs and eyes which acts to diffuse the pulsating stimulation, enabling the user to experience the session with eyes open. Recently, clear eye frames have been produced with low intensity LED's that enable the user to experience light and sound sessions with eyes open, enabling the user to simultaneously experience the session while studying, working at computer workstations or during beta EEG/neurofeedback training. Using red LED's are for maximizing session experiences. Other colors create a softer effect.

Light & Sound Designer is a software package that allows sessions to be designed in a spreadsheet format on a PC computer. Graphically displayed and stored on disk format, this software is then downloaded into all Photosonix downloadable systems.

Light & Sound Librarian helps organize sessions, sets up downloads for multiple sessions and prints session lists for all Photosonix downloadable systems.

Lucid dreaming refers to being aware that you are dreaming while you are experiencing your dreams.

Mind Machines are devices that entrain your mind to allow you access to different states of awareness. Utilizing light and sound for stimulation, these instruments are mainly used for relaxation and learning.

Musical scale is how the sound frequencies are chosen as notes for playing music.

Neurofeedback incorporates EEG instrumentation with biofeedback to heighten the participant's ability for achieving targeted brainwave frequencies.

Octave is an interval between notes that has a frequency ratio of 2/1. The tempered scale has 12 total notes, seven natural notes and five sharps and flats.

Offset frequency occurs with dual binaural beats. It is the frequency the mind actually hears when stimulated by two different frequencies. For example, the mind is stimulated by a frequency of 200Hz and simultaneously by 190Hz. The mind subtracts the difference, which is 10Hz. This is the frequency actually heard by the mind and is a frequency associated with the alpha brainwave state.

Phase is used for timing the relationship between light and sound stimulation- between the eye frames and headphones.

Pitch is the frequency of sound used for auditory stimulation and is thought of in terms of bass (the low notes) through soprano (the high notes).

Ramping is a gradual method of changing (either increasing or decreasing) the parameters of light and sound stimulation when designing a session.

Schumann Resonance Frequency is 7.83Hz and is referred to as the frequency of the earth.

Sessions are built-in, downloadable or custom made sequences of changing light and sound stimulation for specific purposes, including relaxation, alertness and learning enhancement.

Sine wave sound are pure tones consisting of only one frequency.

Soft on/off refers to beginning a session by gradually increasing (ramping) the light and sound stimulation and ending a session by gradually decreasing the light and sound stimulation. This enables the transition into and out of a session to be much more pleasant, without any sudden jerks.

Tactile stimulation are specifically designed frequencies that produce physical sensations. When utilized with light and sound, this experience creates a complete "mind/body" experience.

Tempered scale is a musical scale utilized in modern western music.

Theta brainwave frequencies are associated with states of deep relaxation, rapid learning and creativity. These frequencies range between 4-7Hz.

Thwack is a little known technique that utilizes gamma light, sound and tactile stimulation, along with motion, to achieve an immediate state of extreme alertness and motivation. Created by David McLaughlin, we dare not say this approach has any medical or theraputic value, but possesses immense entertainment value. However, our research indicates it has theraputic potential and does assist in achieving maximum performance in athletic endeavors.

Tone is the kind of sound stimulation used in light and sound sessions. Examples of tone are: pulsed tone, pulsed surf, alternating tone/surf, pulsed chord, binaural beats, binaural beats with a metronome tick at a specific frequency, dual binaural beats and dual binaural beats with surf.

Visuals are the wavy, shifting color patterns perceived when observing the pulsating lights during a light and sound session. These patterns are the result of the interaction of the light pulse rate of your instrument to the sensing/processing rate of your eyes.

Volume refers to the loudness of the sound stimulation.